Children, Youth and Development

WITHDRAWN

Nearly half of the world's population is under the age of twenty-five and almost 90 per cent of these young people live in Third World countries. The experiences of children and youth in the Third World are extremely diverse. They are affected by global processes, such as development and globalisation, and international agreements and policies relating to children and youth, as well as by more local contexts.

Children, Youth and Development focuses on the ways in which changes in development theory and practice have impacted on young people's lives and considers the influence of the differing geographical and social contexts in which young people live. This text addresses issues that preoccupy children and youth, or dominate policies towards them, with topics as diverse as health, education, child labour, youth unemployment, street children and child soldiers. Finally, the book examines how children and youth, as bearers of rights and with the enthusiasm and energy to bring about change, can be enabled to participate in 'development'.

This accessible textbook contains numerous student-friendly features, including boxed case studies from across the Third World, summaries of key ideas, discussion questions and annotated guides to further resources on each chapter's main topics, as well as diagrams, tables, maps and photographs.

Nicola Ansell is a Lecturer in the Department of Geography and Earth Science at Brunel University.

Routledge Perspectives on Development

Series Editor: Tony Binns, *University of Sussex*

The Perspectives on Development series will provide an invaluable, up-to-date and refreshing approach to key development issues for academics and students working in the field of development, in disciplines such as anthropology, economics, geography, international relations, politics and sociology. The series will also be of particular interest to those working in interdisciplinary fields, such as area studies (African, Asian and Latin American Studies), development studies, rural and urban studies, travel and tourism.

Published:

David W. Drakakis-Smith
Third World Cities, Second Edition

Jennifer A. Elliott
An Introduction to Sustainable Development,
Second Edition

Nicola Ansell
Children, Youth and Development

Janet Henshall Momsen
Gender and Development

Kenneth Lynch
*Rural–Urban Interaction in the Developing
World*

Katie Willis
Theories of Development

Forthcoming:

Hazel Barrett
Health and Development

Alison Lewis and Martin Elliott-White
Tourism and Development

John Soussan and Mathew Chadwick
Water and Development

Chris Barrow
*Environmental Management and
Development*

Tony Binns, Peter Illgner and
Etienne Nel
Indigenous Knowledge and Development

Children, Youth and Development

Nicola Ansell

LONDON AND NEW YORK

First published 2005
by Routledge
2 Park Square, Milton Park, Abingdon, Oxon OX14 4RN

Simultaneously published in the USA and Canada
by Routledge
711 Third Avenue, New York, NY 10017

Routledge is an imprint of the Taylor & Francis Group, an informa business

Typeset in Times by
Florence Production Ltd, Stoodleigh, Devon
Printed and bound in Great Britain by
TJI International, Padstow, Cornwall

British Library Cataloguing in Publication Data
A catalogue record for this book is available from
the British Library

Library of Congress Cataloging in Publication Data
A catalog record of this book has been requested

ISBN 10: 0–415–28768–5 (hbk)
ISBN 10: 0–415–28769–3 (pbk)

ISBN 13: 978–0–415–28768–5 (hbk)
ISBN 13: 978–0–415–28769–2 (pbk)

To Mum and Dad,
with thanks for my childhood
and youth

Contents

Plates

 **Figures**

 **Maps**

Tables

Boxes

Acknowledgements

I am grateful to many people who have been supportive in producing this book. First, I am indebted to Elsbeth Robson, who first got me interested in the geographies of young people in the Third World, Lorraine van Blerk, for countless discussions about children and development, and the staff and students in the Children's Geographies Research Unit at Brunel University who have challenged me to think about young people's lives in new ways. Thanks, also, to my colleagues in the Department of Geography and Earth Sciences at Brunel for continuing to make this such a conducive and enjoyable place to work.

At Routledge, I would like to thank Andrew Mould and a succession of editorial assistants (Ann Michaels, Melanie Attridge and Anna Somerville) for their help in the course of writing this volume. I am also grateful to colleagues, friends and family who have kindly allowed me to use their photographs. Lastly, the book could not have been written were it not for the young people in Lesotho, Zimbabwe and Malawi who have taught me so much about their lives.

Abbreviations

ACRWC	African Charter on the Rights and Welfare of the Child
AIDS	Acquired Immune Deficiency Syndrome
ARI	acute respiratory infection
CBR	community-based rehabilitation
CCCS	Centre for Contemporary Cultural Studies
CEDC	children in especially difficult circumstances
COSC	Cambridge Overseas School Certificate
CRC	Convention on the Rights of the Child
DAP	developmentally appropriate practice
DPO	disabled people's organisation
ECD	early childhood development and care
EFA	Education for All
EPI	Expanded Programme on Immunisation
EWP	Education with Production
GATS	General Agreement on Trade in Services
GAVI	Global Alliance for Vaccines and Immunisation
GDP	Gross Domestic Product
GER	gross enrolment ratio
GNP	Gross National Product
GOBI	growth monitoring, oral rehydration therapy, breast feeding and immunisation
GUIC	Growing Up in Cities
HIPC	Heavily Indebted Poor Country
HIV	Human Immunodeficiency Virus
ICDS	Integrated Child Development Services
IDP	internally displaced persons

IFI	international financial institution
ILO	International Labour Organisation
IMCI	integrated management of childhood illness
IMF	International Monetary Fund
INGO	international non-governmental organisation
IPEC	International Programme on the Elimination of Child Labour
KAP	knowledge, attitudes and practices
LDC	Least Developed Country
NEPAD	New Partnership for Africa's Development
NGO	non-governmental organisation
OPEC	Organisation of Petroleum Exporting Countries
ORS	oral rehydration salts
ORT	oral rehydration therapy
RORE	rate of return to education
SAP	structural adjustment policy
STI	sexually transmitted infection
TRIM	Trader-Related Investment Measures
TRIP	Trade-Related Intellectual Property Rights
UN	United Nations
UNESCO	United Nations Educational, Scientific and Cultural Organization
UNFPA	United Nations Population Fund
UNICEF	United Nations Children's Fund
UPE	universal primary education
USAID	United States Agency for International Development
WHO	World Health Organisation
WSC	World Summit for Children
WTO	World Trade Organisation

 # Introduction

This book is concerned with the experiences of children and youth in the Third World, and in particular the ways in which young people are affected by global processes including development, globalisation and the effects of international agreements and policies relating to children and youth. Given that children, youth, development and the Third World are all contested terms, lacking precise or agreed definitions, the book's first task is to outline how and why these terms are used.

There is no universal definition of 'child' or 'youth', let alone agreement about what it means to be a child or a youth. This book broadly concerns the young people that fall within the United Nations (UN) definitions, where 'children' are aged 0–17 and 'youth' aged 15–24. These definitions, based upon chronological age, do not correspond to those in use in all countries. In Brazil, for instance, the age of majority is 21 and Malaysia's Youth Council defines youth as those aged 15–40. Furthermore, most countries operate multiple age thresholds according young people different statuses en route to full adulthood. Nor is chronological age a universally accepted basis on which to determine who is, or is not, a child or a youth. Other factors, including whether or not an individual has been through an initiation ritual, has married or has borne children are more meaningful in many contexts.

'Development' is an equally problematic notion, employed variously to describe a set of inevitable economic and social processes, a desired endpoint for all societies, or a range of interventions aimed at improving life in poor countries. The term is used here to signify policies and practices promoted by Western-dominated institutions for implementation in parts of the world characterised by relatively high levels of poverty, with the intention of improving quality of life. Development is promoted across Africa, Latin America, the Caribbean and many countries in Asia. Most of these countries were once subject to European colonialism, and some, more recently, to the influence of the USSR. While widely represented in positive terms, the specific policies, premises and even the overall goals of development are increasingly questioned.

Throughout the book, the regions that are subject to 'development' are referred to as 'Third World'. Although far from satisfactory, the term 'Third World' is used here in preference to others that are in common use. 'South' and 'Global South' are geographically inaccurate and suggest a symmetry that depoliticises differences between rich and poor worlds. To speak of countries as 'developing' or 'underdeveloped' identifies them only in relation to an implicitly agreed normative process of 'development'. Coupled with 'developed world', 'developing world' also implies economic convergence, when in reality the world is becoming more unequal. It must be recognised, nonetheless, that any term employed to impose a common identity on such diverse parts of the world is inadequate, not only in assigning a false homogeneity to Third World countries, but also falsely implying homogeneity of experience among young people across these parts of the globe.

Why a book on children, youth and development?

Children and youth, as defined by the UN (and grouped together here as 'young people'), account for nearly half the world's population (Table 1). Whereas in the West, the number of young people is falling, both in absolute terms and relative to the population as a whole, globally the proportion of the population under the age of 25 has risen steadily over the past century and is set to continue growing. This growth is accounted for by demographic change in Third World countries, where high fertility rates have swelled the younger populations. In some African countries more than half the population are under 18. Somalia, for instance, has a typical pyramid-shaped population distribution, with 54.6 per cent of its people under 18 (United Nations 2001b) (see Figure 1). Furthermore, while two-thirds of the world's total population inhabits Third World

Table 1 *Population estimates for 2000*

	Aged 0–17		Aged 15–24		Aged 0–24	
	Thousands	*% of total population*	*Thousands*	*% of total population*	*Thousands*	*% of total population*
World	2,104,666	35.6	1,062,283	17.5	2,876,808	47.5
Africa	392,277	49.4	160,925	20.3	499,117	62.9
Asia	1,318,969	35.9	652,137	17.8	762,700	48.0
Latin America and the Caribbean	195,075	37.6	101,323	19.5	264,883	51.0
'Less developed regions'	1,886,722	38.8	899,816	18.5	2,496,397	51.3

Source: Based on United Nations (2001b).

countries, the Third World accounts for 90 per cent of 0–17-year-olds and 85 per cent of 15–24-year-olds (United Nations 2001b).

Given these statistics, a book about children and youth in the Third World is a book about over 40 per cent of the world's population. Despite the high numbers, until relatively recently children and youth received little attention from either academics or policy-makers, a situation that has begun to change over the past two decades. Although the sheer numbers involved give weight to the importance of considering the lives of young people in the Third World, they

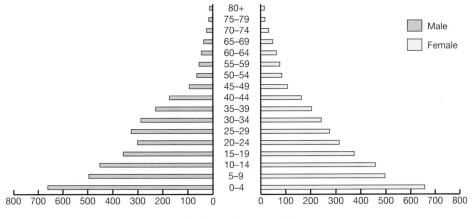

Population (in thousands)

Figure 1 *Population pyramid for Somalia.*
Source: Data from US Census Bureau (2004).

also raise the question of whether there is sufficient to unite these individuals to justify a book about them.

Much research and policy making narrows its scope by treating children and youth as separate categories, and focusing on either one or the other. This is not the approach taken in this book, for a number of reasons. First, neither children nor youth constitutes a homogeneous group, any more than children and youth together. While a 24-year-old has little in common with a three-month-old baby, a 17-year-old shares little more. The only thing that binds 0–17-year-olds together, and divides them from those aged 18 or over, is their lack of full legal standing, and even this is not a clear-cut division: some legal rights are acquired at other ages in most societies, and in some 18 is not the age of majority. Furthermore, defining all young people under 18 as 'children' risks infantilising those towards the upper limit, and does not adequately address their needs. Even the UN finds its own two categories inadequate, choosing different terms to relate to young people of different ages, in order to address the very different issues they face (Figure 2).

The emphasis on two distinct categories has also led to an unhelpful ghettoisation within research and professional communities (Caputo 1995). Even though 15–17-year-olds, in particular, fall into both categories, studies of youth focus on different aspects of life from studies of children. In Latin America, for instance, youth have mainly been studied by social scientists in relation to sexual behaviour, pregnancy, drugs and violence (Welti 2002). Young people defined as youth are thought of in very different ways from those defined as children, in both academic and political contexts. Indeed, these conceptual distinctions are used by young people and their supporters: teenagers sometimes claim the label 'child' to win sympathy, whereas those who wish to denigrate them may call them 'youth' (Boyden 2000). Bringing children and

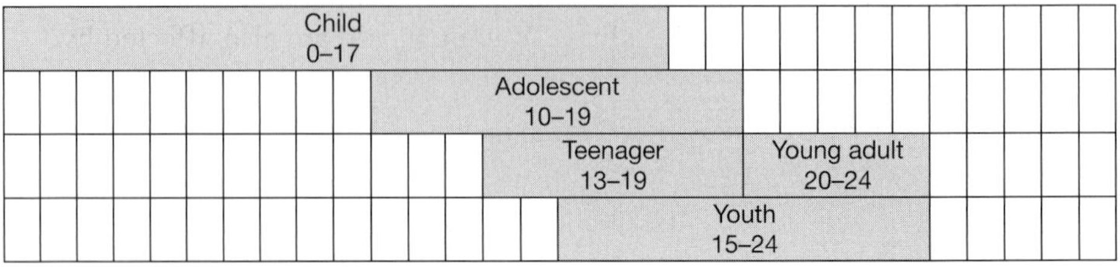

Child 0–17		
Adolescent 10–19		
Teenager 13–19	Young adult 20–24	
Youth 15–24		

Figure 2 *Terms in use by UN agencies.*

youth together allows us to question some of the conventional ways of thinking about either category, as well as serving to disrupt the adult–child conceptual binary (Aitken 2001).

Although childhood and youth cover a range of ages, there is some commonality of experience insofar as young people everywhere experience a (varying) degree of age-based discrimination, and in most parts of the world it is felt that young people need to be formally provided with an education. The health care needs of young children differ from those of adults, and children are also generally the first to die in situations of famine. Childhood and youth are times of shifting identities (Aitken 2001), and in research, children and young people often report feeling powerless and excluded (Roche 1999). However, besides being fractured by differences of age (and age-related status), children and young people are riven by many other lines of difference. These range in scale from individual bodies (sex, dis/ability, etc.) through to broad societal differences between young people growing up in different parts of the world. In many cases these other aspects of young people's identities and experiences are more important to them than those relating specifically to age.

What unites the young people whose lives are considered in this book is not so much that in all societies there are particular issues that affect the young in different ways from the older populations, as the fact that young people in poor countries across the world are affected by particular discourses, practices and policies that are international in scope and character. Young people in Third World countries are affected by 'development' – a process imagined and initiated by the West and implemented through non-governmental organisations (NGOs) and governments in poor countries. International organisations and meetings are also at the centre of the export of Western ideas about childhood and youth to other parts of the world. Both discourses of development and 'global' models of childhood and youth are the basis for interventions at local, national and international levels. Some of these are aimed specifically at young people. Others do not take young people into consideration at all, but impinge on their lives. Young people are also affected by wider global cultural processes. Such processes do not have uniform effects on all young people, but interact with local conditions.

It is important, then, to be conscious that although children and youth in the Third World are affected in various ways by global processes, their lives remain very diverse. Terms such as 'the child', 'the girl child' or 'the African child' are best avoided, because they suggest there is a single 'normal' childhood that could be experienced by all children, girls or African children. No stereotype can capture the real lives of all children (Boyden and Ennew 1997).

In examining the lives of young people affected by development, there are three other principles that inform the approach taken in this book. First, the focus of the book is not exclusively on young people, as young people's lives are lived within social contexts. Most young people belong to families, communities and nations, hence it is necessary to examine how they experience their relationships with others, and how adults conceptualise and influence the lives of children and youth (Boyden and Ennew 1997). Second, children and youth are symbolically connected to the future – they represent potential (Stephens 1994). Societies tend to think about children and youth in relation to the future – both their own futures and the future of the society more widely. However, young people live in the present. It is important not to regard children as 'human becomings', denigrating their present lives, while focusing only on the adults they will become, yet the future-orientation of childhood is also significant. Failure to give attention to young people's lives today, through, for instance, investment in their health and education, will affect them not just today, but possibly throughout their lives. Last, the book takes the view that children and youth are actors in their own lives, not merely objects of development or victims of history. All young people exercise some control over their own situations, and this should be recognised both by policy-makers and practitioners and by those studying young people's lives.

Organisation of the book

The first three chapters of the book focus on the broad contexts of young people's lives. Chapter 1 examines Western discourses of childhood and youth and the ways these have influenced the emergence of policies and practices at the international level. The second chapter considers 'development' as a set of processes that has been implemented in the Third World over the past half-century and the differential impacts this has had on children and youth. Chapter 3 addresses the social and cultural contexts of young people's lives, focusing particularly on the diversity of their experiences.

The next four chapters concern aspects of children's lives that have received particular attention in the international arena and in development policies. Health, education and work (Chapters 4, 5 and 6) impinge on the lives of most young people (if only through their absence). Chapter 7 examines the lives of young people who are considered to be living in especially difficult circumstances. Chapter 8 looks at rights and participation – approaches to improving children's life experiences – and the power relations that inhibit such improvements. The book concludes with a short postscript that brings together some of the key themes.

This apparently neat division of the book into separate chapters hides multiple interconnections between the issues considered in each. Children in poor health, for instance, receive less education and gain less from the education they receive. Equally, educated youth are often better able to secure their own health needs. Attendance at school may preclude participation in paid employment – and vice versa, although many young people combine the two. Those who do not attend school are less likely to find formal sector employment as youth. Children's work includes forms of labour that qualify as 'especially difficult circumstances', such as soldiering and sex work. Children who work or live in difficult circumstances may, as a consequence, suffer poor health.

Many of the issues covered are controversial, and the intention of the book is to promote debate and further examination of these issues, rather than to provide definitive answers. To this end, each chapter concludes with a list of the key ideas conveyed in the chapter, followed by three or four questions for discussion. Also provided are suggestions of resources that might be consulted for further information relating to the themes of the chapter.

1 Global models of childhood and youth

Key themes

- Popular Western constructions of childhood and youth
- Academic theories of childhood and youth
- Exporting global models of childhood and youth.

Childhood and youth are understood in very different ways in different societies and at different times. This was brought sharply to our attention by the historian Aries (1962), who argued that, in the West, the notion of childhood did not exist prior to the fifteenth century: once the physical dependency of infancy was ended, children were treated as miniature adults, fully integrated into social and economic life. In the sixteenth century differentiation between childhood and adulthood became more distinct, but only in the late nineteenth century, as schooling became universal and children were progressively removed from adult society into their own spaces, did many of today's popular ideas about childhood emerge. Although this thesis has been critiqued, it is widely accepted that medieval concepts of childhood differed greatly from those prevailing today (Heywood 2001), and conceptualisations of childhood continue to change. Even within the UK, expectations concerning the treatment and behaviour of children today differ from those common in the 1950s.

Not only do notions of childhood and youth change over time, there are also considerable differences between societies. Margaret Mead's

(1928) research suggested that although young people in Samoa experienced adolescence as a life phase, it was very different from the time of 'storm and stress' believed to characterise youth in the West. Even within societies there are considerable differences in what is expected of children between different social classes and ethnic groups, and between girls and boys. Childhood and youth, then, are not natural, universal categories, and the difference between children and adults in a given society cannot simply be read off from physical difference. Movement from one status to another is not simply a matter of physical change (Jenks 1996). Rather, '[t]he category child [is] inscripted on the body through a lengthy historical process' (Holloway and Valentine 2000: 4).

As conceptualisations of childhood vary over time and space, so too do experiences of being a young person. These two aspects are, of course, related. Expectations concerning the part children *should* play in society both reflect and influence the roles they *do* play. Social constructions of childhood and youth are both products of historical circumstances and have real effects on young people. In studying the social construction of childhood, it is important to avoid 'abandoning the embodied material child' (James *et al.* 1998: 28).

While childhood and youth are not natural physical categories, there are biological and physiological facts that constrain and shape children's lives, particularly in early childhood. Aspects of infancy, including the acquisition of language and mobility skills, are almost universal. 'Childhood integrates biological and social processes' (Scheper-Hughes and Sargent 1998: 1), but the social contexts of children's everyday lives are crucial in shaping their experiences.

The diversity of both concepts and experiences of childhood and youth will be explored further in Chapter 3. This chapter examines the origins of Western conceptualisations of childhood and youth which, while not universally subscribed to, have spread across the world. It then considers the processes through which these 'global models' have been exported from the West.

Popular Western constructions of childhood and youth

Constructions of childhood and youth encompass both normative and descriptive elements: the characteristics attributed to young people of different ages and sexes, how they are treated, the ways they are expected to behave, how they are valued by society and what is considered desirable in a young person. Western ideas about childhood and youth have shaped policies and practices, not only in the West, but also in former colonies of the West, in the international arena and in development interventions. There is, however, no singular coherent Western conceptualisation of either childhood or youth, but rather a number of discourses that are often seemingly contradictory and have their origins in different historical contexts.

Childhood

Particular conceptualisations of childhood need to be understood in relation to the social conditions that gave rise to them (Heywood 2001). For instance, when many children died at an early age, less investment was made in young children, either economically or socially. Indeed, infant mortality was first calculated in the late nineteenth century, although the data this required had long been available (Aitken 2001). Older children were accorded greater value, in part for the economic contributions they made to impoverished households. In the West today, in contrast, '[c]hildren have become relatively worthless economically to their parents, but priceless in terms of their psychological worth' (Scheper-Hughes and Sargent 1998: 12). Social conditions do not arise from a vacuum, however, but relate to wider economic and political processes. It has been argued that hardening of the child/adult dichotomy was central to the development of modern capitalism and modern nation states (Stephens 1995).

Although concepts of childhood may be related to prevailing conditions, there have long been ambiguities in the ways in which children are understood in Western thought. To a large extent, concepts persist beyond the conditions giving rise to them, and Heywood (2001: 20) urges us to 'think in terms of . . . competing conceptions of childhood in any given society'. Over the past four centuries in the West there have been two predominant yet contradictory images of children, characterised by Jenks (1996) as 'Apollonian' and 'Dionysian' views (Figure 1.1). Although both views have long roots, it is argued that the Dionysian view dominated in the West prior to the twentieth century, but that increasingly the Apollonian image has come to the fore, particularly influencing society's views concerning younger children. Both are somewhat problematic (Holloway and Valentine 2000) and oversimplify sets of disparate ideas.

Dionysian	Apollonian
Children should be seen and not heard	Childhood is a time for play, and not for work
Children need protection from themselves	Children need protection from the world
Childhood is a time to learn discipline	Children are innocent
	Children are passive
	Childhood should be happy

Both
Childhood is a time set apart from the adult world
Children belong in families
Children are closer to nature than adults
Children are incomplete – less than adult

Figure 1.1 *Western concepts of childhood: legacies of the Dionysian and Apollonian views.*

The Dionysian view perceives children as 'little devils', born into the world with, according to Catholic doctrine, original sin, and in need of strict moral guidance (Jenks 1996). Children are seen as impish pleasure seekers who are easily corruptible and may be led into bad habits unless they are kept occupied and strictly disciplined. Discipline was needed to save children from themselves: amid high infant mortality, the priority was to ensure children's redemption and a place in heaven (Holloway and Valentine 2000). From the Puritan seventeenth century to the childhoods described by Dickens in the nineteenth century, the Dionysian view inspired injunctions not to 'spare the rod and spoil the child'. Even today, the idea that children should be kept away from certain places and protected from 'dysfunctional families' is fed by the idea that such conditions may serve to liberate the demonic forces already present in the child (James *et al.* 1998). Allowing the evil forces inherent in children to be unleashed would threaten not only children themselves, but also wider adult society (James *et al.* 1998). Fear of the 'polluting' effect of disruptive and anti-social children informed social policy in Victorian Britain and, Boyden (1990) argues, it influences policy responses concerning children in the Third World today.

By contrast, the Apollonian view casts children as 'little angels', born innocent and untainted by the world. Rather than needing to be beaten into submission, the Apollonian child is naturally virtuous and needs encouragement and support (Jenks 1996). Children are viewed as special; needing protection, not from their own inner weakness, but because their intrinsic natural state is desirable and worth defending (James *et al.* 1998: 210). While this perspective is apparent in the eighteenth-century writings of Rousseau, particularly

Émile (James *et al.* 1998), its growing prevalence in the West is often characterised as a middle-class sentimentalisation of childhood, partly rooted in adult nostalgia (Boyden 1990). It, too, has had practical implications. The Apollonian view sees children as individuals, more than as a collective future adult society, and has come to dominate Western ideas about how children should be treated, both by society in general, and within specific contexts of, for example, home and school (Jenks 1996).

These twin perspectives on children can be seen to underlie many of the characteristics popularly attributed to childhood. The idea that childhood should be a time set apart from the adult world is rooted in an Apollonian notion of children as different and special. Although this idea began to emerge among the wealthy in eighteenth-century Europe (Aries 1962), even in the nineteenth century there were many who believed children should be occupied in hard labour for their own good and that of society: that children's natural idleness should be disciplined. The Western view that childhood should be happy, a time devoted to play and learning rather than work, is thus relatively recent.

While the Apollonian and Dionysian perspectives are clearly contradictory, they are not wholly unrelated. Both see children as closer to nature than adults: either as 'savage' and in need of taming, or as pure and in need of nurture (Jenks 1996). Both Dionysian and Apollonian views of childhood consider children fundamentally different from adults. Aries (1962) pointed to the increasing physical segregation of children from adults. Due to both Dionysian and Apollonian fears, children have been progressively removed from 'adult' public space, retreating to either designated children's spaces or the home (Holloway and Valentine 2000). Equally, children are seldom accorded a public voice. The notion that children should be 'seen and not heard' illustrates adult society's disregard for children's opinions. Children have come to be seen as incomplete – passive recipients of adult care and tuition, rather than agents in their own lives, let alone in wider society. They belong to families, and it is their families that act upon their behalf and represent their interests.

Children in the West were described in 1975 as 'being seen by older people as a mixture of expensive nuisance, slave, and super-pet' (Holt 1975, cited in James *et al.* 1998: 210). By the start of the twenty-first century, Western childhoods are described in predominantly Apollonian terms. Victorian childhoods are now regarded as excessively harsh and society is supposedly more child-centred. However, the media focus on children's innocence and vulnerability bears little relation to the real lives and interests of young people. Subject to an increasingly different set of rules from adults, '[c]hildren can no longer be routinely mistreated, but neither can they be left to their own devices' (James *et al.* 1998: 14).

A series of discourses describing the needs and characteristics of children have become conventional wisdom. Significantly, these discourses 'begin from a view of childhood outside of or uninformed by the social context within which the child resides' (James *et al.* 1998: 10). Viewing childhood as a natural state has contributed to a tendency to universalise Western concepts (Jenks 1996), assuming they apply equally to non-Western contexts. However, the childhood characteristics that are common currency in the discourse of Western media only partially relate to Western childhoods as lived today (or even in the past). They deviate sharply from the experiences of many children in the Third World. They are nonetheless powerful in their impacts on development policy and practice, and their traces appear in media representations of Third World children. In particular, children's innocence, vulnerability and need for protection are frequently stressed, including by development practitioners in their efforts to raise awareness of the 'plight' of children in the Third World (Boyden 1990).

Youth

The dichotomy drawn between children and adults has left limited space for those who sit inconveniently at the boundary. While the West has had a clear (albeit semi-coherent and shifting) notion of childhood since at least the sixteenth century, the border between childhood and adulthood has been rather mobile and subject to less entrenched discourses of its own. Youth is less closely related to physical distinctiveness than childhood, and is often described as a Western concept. In medieval times, 'adolescents were looked upon with some distaste by clerical figures, on account of their licentiousness and "carnal lust"' (Heywood 2001: 15) – yet the term 'adolescence' was coined only in the early twentieth century. According to Aries a 'quarantine' period between childhood and adulthood emerged in early eighteenth-century Europe (Valentine *et al.* 1998). Later in the eighteenth and nineteenth centuries the middle classes became concerned with the need to control not only their own offspring, but also young working-class men, who were increasingly associated with danger (Finn 2001).

Youth has grown progressively longer in the West, as the gap between childhood and childbearing has extended. Young people seek to distance themselves from the world of children but 'may appear threatening to adults because they transgress the adult/child boundary and appear discrepant in "adult" spaces' (Sibley 1995 in Valentine *et al.* 1998: 5). The predominant view over the past 150 years has been of 'youth-as-trouble', with repeated moral panics. This may relate in part to the fact that 'youth remains a major point of symbolic investment for society as a whole' (McRobbie 1993, cited in Valentine *et al.* 1998: 9). If the 'state of youth' is seen as the signifier of the 'state of the nation' it is unsurprising that society is concerned about troublesome youth (Griffin 2001). Youth is also understood as a time of transition – most importantly from school to work. Youth unemployment is a perennial concern of Western governments. In contrast to children, youth are seen as political – both potential (and threatening) political actors themselves and in need of political 'remedies'.

Youth are not seen in wholly negative terms. The 'teenager' was invented in the West in the 1950s, a time of relative affluence, as a new variety of consumer with disposable income and few responsibilities. Berger (1971) described young Americans as characteristically 'spontaneous, energetic, exploratory, venturesome, vivacious, disrespectful, playful and erotic' (cited in Wulff 1995: 7). These descriptors apply not only to teenagers, and do not describe all teenagers – indicative of the impossibility of arriving at a universal definition of youth, even within a particular society. Unlike 'youth-as-threat', however, 'youth-as-fun' carries an assumption that young people need not be taken seriously.

Academic theories of childhood and youth

Popular Western concepts of childhood and youth have shaped the interests and assumptions of those studying young people and the contexts within which they are able to do so. Both children and youth have been widely neglected in the social sciences, perhaps because until recently the social sciences were male dominated (Prout and James 1990); perhaps because it is assumed that as children's lives are impermanent they are unimportant (Montgomery 2001); or because youth need never be taken very seriously (Wulff 1995). It is significant that studies of children and youth have emerged from distinct disciplines. This has influenced how they have been studied and represented, and also the aspects of young people's lives that research has focused on, forcing them apart in a way that is no longer helpful.

Until recently the area of social science that dominated research with children was developmental psychology. The study of children was initially conceived as a way of finding solutions to general psychological problems, rather than concerned specifically with children or even child development (Jenks 1996). Educationalists were concerned with children as already-constituted recipients of schooling, but seldom delved further into children's lives. Within anthropology, early work with children by some female anthropologists such as Margaret Mead was little recognised within an emerging male-dominated canon (Helleiner 1999). In the 1990s, however, sociologists, anthropologists and geographers began to recognise children as legitimate subjects of study. A new paradigm, the 'new social studies of childhood', has stimulated research that is critical of the psychological perspective.

Whereas the study of children has its roots in the fields of education and psychology, the study of youth is more recent, and began in the 1950s and 1960s among criminologists, and later psychologists and sociologists, inspired by concerns surrounding the 'nuisance' posed by working-class adolescents on urban streets (Valentine *et al.* 1998). In recent years anthropologists and geographers have begun to take a wider interest in the lives of 'ordinary' (i.e. untroublesome) youth.

Psychological perspectives

Child development

Developmental psychology accepts two assumptions rooted in popular Western concepts of childhood: children are natural rather than social phenomena, and their maturation is a natural and inevitable process (James *et al.* 1998). It also focuses exclusively on the cognitive.

The most influential developmental psychologist of the twentieth century was Jean Piaget. Piaget's model of the naturally developing

Table 1.1 *Piaget's developmental stages*

Age	Stage	Competencies
0–1½	'Sensori-motor'	Knowledge of environment through action and sensory information; begins to understand causality and develop internal mental representations
2–4	'Pre-operational thought'	Development of language; egocentric – cannot take the perspective of another person
5–6	'Intuitive thought'	Begins to order, classify and quantify objects; intuitive problem-solving but thinking remains egocentric
7–11	'Concrete operations'	More complex thought processes, but remains tied to concrete experience; begins to be able to look at things from the perspective of others
12–15	'Formal operations'	Abstract reasoning begins; can think reflexively

Source: Based on Piaget (1972).

child was empirically based and positivistic (James *et al.* 1998). In experiments with young children, Piaget identified a series of stages of cognitive development. According to his model, all children should move through these stages in a universal temporal order that is programmed into their genetic makeup (Table 1.1). The stages are taken as hierarchical with greater status and value attached to the later stages.

According to Piaget, children move from one stage to the next by acting upon stimuli, but they do so in pre-ordained ways. The destination, too, is pre-determined: the ideal is to achieve scientific rationality, to see the world as a set of facts, overcoming the influence of emotion (Jenks 1996). As a hierarchical model, it provides analytic grounds to differentiate between children and adults, justifying the supremacy of adulthood, and it 'further ensures that childhood must, of necessity, be viewed as an inadequate precursor to the real state of human being, namely being "grown up"' (James *et al.* 1998: 18).

Piaget's developmental model is inherently Eurocentric. The ideal of adult cognitive competence is a Western one, based on being able to think about the world using concepts of Western logic. It is also empirically unreliable: children possess some competencies long before Piaget says they do, and in different social contexts different patterns are observed. Piaget's approach has, nonetheless, had huge impacts on education, medicine and government agencies, all of which have been involved in testing children to see whether they match up to 'normal'.

An alternative model of cognitive development is offered by Russian psychologist Lev Vygotsky. In Vygotsky's view, children actively appropriate society, rather than passively growing into it. Children learn to reason through socially structured engagement with those who are already proficient (Modell 2000). They appropriate tools including concepts (language, mathematical reasoning, etc.) and materials from their immediate contexts, initially putting them into practice in interaction with others within that context, but subsequently applying them in other situations (Modell 2000). Because children appropriate from their contexts, the context is much more important than for Piaget, and children's development needs to be understood in relation to history. As contexts change, they affect what children take from them and also affect the way that children's appropriations themselves modify the contexts that guide their development (Modell 2000). As with Piaget, however, the goal of cognitive development is to transform the child into a rational and complete adult (Caputo 1995).

Attachment theory

Similarly influential has been the work of John Bowlby (1966/1951), who argued that children should not be 'deprived' of contact with their mothers (or 'permanent mother substitute') during the first three years of life, as this would interfere with the formation of an attachment relationship that he considered crucial to subsequent social and cognitive development. Although valuable in drawing attention to the fact that children have emotional needs as well as physical requirements, Bowlby's theories are now largely discredited (Smith and Cowie 1988). The research on which Bowlby drew was faulty in its design, and failed to recognise the very varied ways in which young children cope with separation from a primary caregiver (Barrett 1998). Furthermore, in many non-Western societies children have traditionally formed attachment relationships with multiple adults and peers, a situation that enabled their mothers to work and gave children greater security in situations where adult death rates were high (Mead 1966). Research found only 5 of 186 non-industrial societies where the mother was the 'almost exclusive' caregiver of infants (Smith and Cowie 1988). Although it is now argued that a continuous exclusive relationship between child and mother might be harmful to both, there remains a strong expectation in Western society that children should follow a trajectory from attachment to detachment (Burman 1995a).

Adolescence

The term 'adolescence' was invented by American psychologist G. Stanley Hall to describe a period of 'storm and stress' observed in young people, mid-way between the dependency of childhood and the mature stability of adulthood. This turmoil was attributed by psychologists to hormonal changes taking place at puberty (Griffin 2001). Adolescence has come to be seen as a period when young people are particularly at risk: a biologically determined stage of vulnerability to social vices (Finn 2001). Social work in the early 1900s began to focus on the individual as the locus of concern and framed delinquency in pathological terms: boys were affected by stubbornness, disorderly conduct and mental slowness, while delinquency in girls was linked to their sexuality (Finn 2001). Concern was stimulated in part by the needs of capitalism for amenable workers, hence social workers were sent into homes to instruct parents on childrearing (Finn 2001).

The groups pathologised as 'in trouble', or 'troubling' or 'at risk' in the West today are marked by certain social attributes. Those 'in trouble' are mainly working class, male and/or ethnic minority; those who are 'troubling' or 'at risk' are predominantly female, working class and/or from ethnic minorities (Griffin 2001). Minimal attention is given to the contexts in which young people live.

> In the early 1980s, a host of new psychiatric diagnostic categories, such as 'oppositional defiant disorder' (defined by losing one's temper, arguing with adults, defying authority, deliberately doing things that annoy others, blaming others, being overly sensitive and angry and swearing a lot), were officially recognized as categories by which to define stubborn and troubling young people in terms of disease.
>
> (Finn 2001: 177)

The behaviour of some youth was attributed to individual biomedical causes, increasingly seen as amenable to individual treatment using drugs. The assumption that youth everywhere are characterised by biologically determined 'storm and stress' has, however, been widely critiqued, beginning with the work of Margaret Mead, who found that girls in Samoa did not experience adolescence as a turbulent time (Finn 2001).

Approaches from sociology, anthropology and cultural studies

Socialisation of children

Developmental psychology is not the only discipline that has sought to explain how 'immature, irrational, incompetent, asocial [and] acultural' children are transformed into 'mature, rational, competent, social and autonomous' adults (Prout and James 1990: 13, after Mackay 1973). Sociologists have been more conscious of the diversity of childrearing practices, and have suggested that, rather than developing 'naturally', children are socialised into different societies through the different ways societies treat them. In Talcott Parsons' (1951) structural functionalist world-view, socialisation is a mechanical process in which children respond to external stimuli provided by adults. Children passively receive and reproduce the culture of the society in which they grow up, gradually acquiring adult social roles (Caputo 1995). Society shapes the individual, and in the process reproduces itself: there is no role for agency, even among adults, and society may therefore be expected to remain static. Children who do not conform, or become 'deviants', are seen as having failed to become socialised.

Youth subculture

Anthropologists have long shown an interest in rites of passage and cultural practices marking the transition to adult status (Griffin 2001), but not from the perspectives of young people (Wulff 1995). In the 1950s, sociologists began to recognise that youth were not simply enacting part of adult culture, but participating in the production of specific youth culture, including through resistance and deviance. Criminologists, psychologists and sociologists became interested in working-class 'gangs' of youths, perceived as delinquent (Valentine *et al.* 1998). Through the concept of subculture in the 1960s, such gangs were studied as 'tribes' (James *et al.* 1998). Subsequent work sought to identify the structural conditions that gave rise to gang cultures. This was formalised in the 1970s by members of Birmingham University's Centre for Contemporary Cultural Studies (CCCS) using Marxian concepts of social class and ideology. Critical of the universalised storm and stress model, British research on youth cultures and subcultures in the 1970s critiqued representations of young people as necessarily 'troubled' or 'troubling' (Griffin 2001).

Using the Gramscian concept of hegemony they sought to explain how social subjects consent to and negotiate their subordination. Working-class white boys were understood as resisting class domination through spectacular forms of style: creating oppositional lifestyles through appropriating and transforming the meanings of mundane objects (Valentine *et al.* 1998). This was, however, a symbolic class war the boys were destined to lose (Wulff 1995).

The work of the CCCS was criticised for neglecting female youth and the home and family as contexts, a situation that began to change in the 1980s under the influence of Angela McRobbie (1978). More recent research has explored the construction of gender among youth and, in the 1990s, ethnicity. Attention has also moved from parochial concerns to the global cultural exchange of commodities and styles, especially the production of hybrid youth subcultures, including in non-Western contexts. No longer are youth subcultures seen as coherent forms of resistance to a coherent dominant culture.

Youth transitions studies

Much recent youth research in the West has been funded by government and NGOs concerned about 'youth-as-trouble' (Wulff 1995). Of particular concern have been young people's transitions to adult life: uncertainties associated with entry to the labour market and leaving home (Griffin 2001). In line with child psychology, young people are expected to move from dependency to autonomy (see Chapter 3 for a critique of this approach), but such transitions are said to be becoming longer and more complex. Research is focused on particular groups (constructed along lines of class, gender, sexuality, race, ethnicity, disability, religion and nationality) and specific problems (teenage pregnancy, youth crime, drug abuse, school dropout and exclusion) – issues that concern governments rather than necessarily young people themselves.

The new social studies of childhood

Increasing concern with agency and context in the social sciences has led to a new approach to childhood studies, and a broadening of interest in children's lives across a wider range of disciplines than in the past (Holloway and Valentine 2000; James *et al.* 1998). The common threads to what are known as the 'new social studies of childhood' are outlined in Box 1.1. This approach is not wholly new:

Box 1.1

Features of the new social studies of childhood

- Childhood is socially constructed;
- Childhood is a variable of social analysis which cannot be entirely separated from other social variables, e.g. gender, class and ethnicity;
- Children's social relationships and cultures are worthy of study in their own right, independent of the concerns of adults;
- Children are actively involved in the construction of their own social lives;
- Ethnography is a particularly useful methodology for the study of childhood as it allows children's voices to be heard;
- The development of a new paradigm is a contribution to the process of reconstructing childhood in society.

Source: Prout and James (1990: 8–9).

---

in the 1970s both Hardman (2001/1973) and Schildkrout (2002/1978) called on anthropologists to study children as people in their own right. Since the 1990s, however, it has become widely accepted that childhood is socially constructed (giving context a greater prominence), and that children are social actors, not simply incomplete beings learning to become adults. On the basis of these premises, James *et al.* (1998) identify four distinct ways of studying children: as socially constructed; as 'tribal', worthy of study independent of adult concerns; as a minority group, subject to discrimination; or as integral to wider social structures.

Studies of social constructions of childhood have highlighted the historical, geographical and social variability of childhood and the moral, cultural and political contexts of assumptions about children (Aitken 2001). Because childhood is not purely a social construct, but lived by real children, questions arise over whether all ways of 'doing childhood' are equally acceptable (Prout and James 1990). It is also worth noting that the new paradigm is itself a normative social construct which conflicts with the way children are seen in many parts of the world.

Studies of children as 'tribal' acknowledge that children are not passive recipients of adult culture, not 'human becomings' but 'human beings' (James and Prout 1990), with culture of their own. Rather than study processes of socialisation, it is appropriate to consider the activities of children in their everyday lives. Children are socially competent, not in terms of having acquired 'adult' competencies through progressive developmental stages, but because they successfully manage interactions with peers and adults and pursue agendas of their own (Hutchby and Moran-Ellis 1998). They are therefore capable of commenting on their own conditions of existence. Furthermore, as socially competent actors, children are not passive and necessarily in need of protection. Indeed, as active agents in their own lives, children are involved in reconstructing childhood itself (Prout and James 1990).

James *et al.*'s (1998) third way of studying children is as a minority group, marginalised like women or ethnic minorities. It is possible to examine how children are discriminated against by society, although there are obvious risks in treating young people as a coherent group. Age-based discrimination exists everywhere, but is expressed in different ways and is not of paramount importance to all children. Furthermore, in contrast to (socially constructed) categories of race, class, dis/ability and gender, the difference of children is different. While every adult is by definition not a child, every adult has been a child and this inevitably influences the way we think about children and childhood. Moreover, children can and generally do become adults (although it is important to recognise that not all children will). Because we were all once young, we presume to understand, which makes it especially important to recognise that there is no truly shared experience of childhood or youth.

Last, studies should recognise that children live within social structures. Particular childhoods are produced through particular institutions, ranging from global political–economic structures to families and communities (Scheper-Hughes and Sargent 1998). Rather than studying children, isolated from their contexts, research is needed that examines how social structures affect children's lives, from their perspectives, and how they, as active agents, shape their societies and their relationships with their families (Montgomery 2001).

While the new social studies of childhood have focused on children, youth have been largely overlooked. The terms 'children', 'youth' and indeed 'adult' are all constructions (Wulff 1995) that would merit deconstruction in relation to the contexts in which they arise. Youth, like children, have been studied as en route to adulthood, more than in their own right. Children and youth are united by the experience of marginality (Wulff 1995), and although youth are not subjected to such extreme silencing and denial of personhood as younger children, their agency has received little attention. In parallel with the growing focus of childhood research on children's everyday lives, youth culture research has begun to move away from the spectacular, deviant, oppositional and marginal to the day-to-day lives of ordinary youth.

The new social studies of childhood are the theoretical paradigm that informs this book. They have also informed much recent thinking in international debate concerning children's lives.

Exporting global models of childhood and youth

Popular middle-class Western notions of childhood and youth, and the Western academic theories they have fed, have not remained confined to the societies in which they originated. The West has actively intervened around the world over several centuries, and taken with it what has been described as a global model of childhood (Box 1.2). In the nineteenth century this Western model was globalised through migration, missionary activity and colonialism. Since the mid-twentieth century it has been development NGOs and international organisations that have exported it to Africa, Asia and Latin America (Ennew 1995). It is important to recognise that the global model of childhood is merely an ideal to which people (are expected to) aspire. For the vast majority of young people in poorer countries (and even in the West), it fails to describe their experiences. Yet, as an ideal, it has had significant material impacts on the lives of young people around the world.

Box 1.2

The global notion of childhood

- There is a natural and universal distinction between children and adults, based on biological and psychological features that are taken for granted;
- Children are smaller and weaker and defined by things they cannot do;
- Children develop through scientifically established stages, for which there is a normal route and timetable;
- The global model is superior to all other childhoods.

Source: Boyden and Ennew (1997).

Colonialism

The attitudes of colonialists to children and young people in the European colonies in the nineteenth century had much in common with attitudes to the children of the poor at home. Prior to the nineteenth century, state and church had intervened little in relation to the condition of children. Foundling hospitals gave sustenance to abandoned infants, but where children remained in the care of families, patriarchal relations remained sovereign. In the nineteenth century a number of societies began to intervene in working-class households, encouraging hygiene and regular family life (Heywood 2001). Concern with children was inspired both by concern about public order and more philanthropic motives, such as underlay the Prevention of Cruelty to Children Act passed in Britain in 1889.

Colonialists and missionaries were also influenced by a widespread idea that colonised people were similar to children, lacking knowledge of the world, irrational and not yet fully human. The 'colonizers (including missionaries) often referred to local people as "children" who needed to be socialized into the customs of Western society' (Lutkehaus 1999: 209). The model of childhood they adopted was in the Dionysian mould:

> the LMS [London Missionary Society], like most missionary societies, assumed that Africans provided a natural focus for Satan's attention, and saw BaTswana [residents of present-day Botswana] as essential sinners. BaTswana were people of wild birth and dark color who, in the order of things, lived like children unknowingly close to evil.
>
> (Landau 1995: xvi)

The function of missionaries was to save the people. If taught literacy, people would be able to read the Bible. Among the BaTswana: 'young and old alike were treated as infants in the "nurseries of education" built by the Nonconformist' (Comaroff and Comaroff 1991: 234).

For some missionaries, however, there was a distinction between adults, who could not be taught, and children who it was possible to save. This inspired some missionary organisations to remove children from their families. Heathen women in India, for instance, were deemed incapable of being 'real' mothers: they were believed not to have their children's welfare at heart (Thorne 1999). Settlements were established for children by Catholic, Anglican and Methodist missionaries in Papua (present-day Papua New Guinea). Kwato

Christian settlement, for example, accommodated about 100 children who were removed from their home villages and submitted to strict discipline. By shielding them from local 'superstitions', the intention was to win lifetime converts, as well as removing children from what were seen as 'heathen habits and ugly practices' in the immediate term. As elsewhere, however, missionaries differed in their views: some opposed the settlements, seeing them as unrealistic and destructive in the way they alienated young people from their culture (Langmore 1989).

International action for children and youth

Children

The twentieth century witnessed a new approach to children overseas, and a number of international events and actions intended to benefit young people (Box 1.3). In the early twentieth century, international concern for children was mainly confined to Europe. Save the Children was set up in London in 1919, to send food to children affected by the blockade against Germany and Austria following the First World War. Save the Children's founder, Eglantyne Jebb, also drafted the Declaration on the Rights of the Child, adopted by the League of Nations in 1924. In 1946, UNICEF was founded, initially with a European remit, although this subsequently expanded (Box 1.4). In 1959, the UN adopted a more comprehensive Declaration of the Rights of the Child. This expressed a relatively paternalistic approach to children, their interests identified entirely by adults, and child welfare identified with family welfare (Boyden 1990).

1979 was declared International Year of the Child. 'Not for the first time the television screens and hoardings of affluent western societies depicted sick and starving children' (James and Prout 1990: 1). The images of Third World children that confronted people in the West represented the antithesis of the Western childhood ideal (Box 1.5). For the first time, however, a notion of 'the World's children' permeated the discourse of UNICEF, the World Health Organisation (WHO) and the International Labour Organisation (ILO) (Aitken 2001) and mobilised a growing commitment to universal children's rights and welfare.

The 1980s were a time of contradiction. While neo-liberal development agendas conceptualised Third World children in an

Box 1.3

International events since 1900

1919	Save the Children founded
1924	League of Nations adopts Declaration on the Rights of the Child (aka World Child Welfare Charter)
1946	UNICEF founded
1959	United Nations adopts Declaration on the Rights of the Child
1979	United Nations International Year of the Child
1985	United Nations International Youth Year: Participation, Development, Peace
1989	United Nations Convention on the Rights of the Child
1990	World Summit for Children
1990	Education for All international conference, Jomtien, Thailand
1996	World Youth Forum, Vienna; Programme of Action for Youth to the Year 2000 and Beyond
1998	World Youth Forum, Braga, Portugal; World Conference of Ministers Responsible for Youth, Lisbon, Portugal; Braga Youth Action Plan
2001	World Youth Forum, Dakar, Senegal; Dakar Youth Empowerment Strategy
2001	Global Movement for Children launched
2002	United Nations Special Session on Children; Global Plan of Action for Children

instrumental way, children also took on the role of 'our most precious resource'. Having in earlier decades been seen as a sensitive barometer of poverty, the first to become sick or die when poverty deepened, in UNICEF's 'child-survival revolution', child survival was an instrument, not a measure, of development. Although critical of neo-liberal development policies, UNICEF was restricted by its major funding source, the US Agency for International Development (USAID), who would fund only technologically simple and cheap forms of intervention (Scheper-Hughes and Sargent 1998).

The other approach to children that emerged in the 1980s centred on children's rights. The UN Convention on the Rights of the Child (CRC) was drawn up by international child-focused NGOs (UNICEF was not initially involved, its energies being focused on child-survival, but joined the process in the late 1980s). It was adopted by the UN in 1989 and came into force the following year

Box 1.4

UNICEF

In the division of Europe following the Second World War, UN assistance to the Soviet-influenced East was ended, but children were seen as innocent victims of a political dispute. An International Children's Emergency Fund (ICEF) was set up to assist children still suffering due to the war, wherever in Europe they came from. UNICEF was launched on 11 December 1946. The UNICEF mandate to support 'child health' allowed it to continue beyond the immediate post-war emergency, and it became a permanent part of the UN in 1953. Retaining its acronym, it was renamed the UN Children's Fund. UNICEF's main concerns in the 1950s were disease eradication and malnutrition.

By the 1960s, UNICEF was arguing that children should be an integral consideration in development policies and plans. A survey was published (UNICEF 1964) reporting on children's needs in relation to health, nutrition, social welfare, work and livelihood. UNICEF began to consider children's intellectual and psychosocial needs as well as their physical welfare, and reached beyond its health and survival mandate to begin to support education. UNICEF also began to give greater attention to the contexts of children's lives – families and communities – rather than seeing children in isolation.

In the 1970s, the perceived failure of Western style development efforts prompted a concern to adopt more appropriate forms of development. UNICEF favoured a basic services approach with 'barefoot' workers delivering local services. The 1980s, however, represented a return to technology and a narrow focus on children. The 'child survival and development revolution' was based on a belief that four low-cost techniques could dramatically reduce infant and child mortality rates. The acronym GOBI referred to growth monitoring, oral rehydration therapy, breastfeeding and immunisation against six vaccine-preventable childhood killer diseases (see Chapter 4). UNICEF also began to take an interest in 'the girl child' and in 'children in especially difficult circumstances' (Chapter 7).

In the 1990s, UNICEF's focus turned to implementing the goals of World Summit for Children (Table 1.2) and children's rights as set out in the United Nations Convention on the Rights of the Child (see Chapter 8). UNICEF's current priorities for children are child protection, immunisation, early childhood, fighting HIV/AIDS and girls' education.

Source: UNICEF (1996, 2003c).

Box 1.5

Media representations of Third World children and youth

> *My name is Alison. I am 17 years old and from Kenya. As a young African,*
> *Kenyan girl it is so frustrating to see the stereotype of the African child and the*
> *African youth in images made by people of particularly the West . . . If I tell a*
> *western person that I don't have flies on my face while my mother carries water*
> *on her head in a desert with lions in the background, most of my Internet*
> *friends are very surprised. When the time comes for me to play a role in the*
> *world, there is none left for me because others' prejudices, backed up by*
> *images they have selected as 'African', have already determined a place for me*
> *without respecting my right to my own image.*
>
> (from the Voices of Youth web project,
> quoted in UNICEF 2002a: 22)

While children are ignored in official statistics, they are represented prolifically in
popular Western culture. Children's voices are seldom heard, but their photographs are
common. There are reasons for employing images of children. Children signify both our
pasts and our futures (Burman 1992). Images of children are used to signify truth, nature,
spontaneity, innocence and dependence. There are elements that do not appear in most
media images of children – qualities of aggression, passion, sexuality and resistance are
expunged 'in order to secure our own sense of adulthood' (Burman 1992: 240). It is
noteworthy that images of children who are poor, sick, disabled, suffering or from the
South fail to conform with the ideal image (Burman 1992). While the children pictured
to represent the future of the planet – to elicit appreciation of their innocence and our
culpability – are white, blond and blue-eyed (Burman 1995a), Third World children are
used in a different way – yet still drawing on the global model of childhood. A BBC
booklet, *All our Children*, that accompanied a TV series coinciding with the CRC,
pictured on its cover a white girl with her arm around a black boy (Burman 1995a).
What did this image convey about the relationship between North and South?

NGOs have been particularly dominant in representing Third World children for a
Western gaze. Such organisations depend for public donations on promoting particular
beliefs about the Third World, and pictures of children help them convey certain
messages. Images of children failing to 'survive' help justify a 'child-saving' mission
(Burman 1995a).

Children are often pictured alone in aid appeals, deliberately separated from adults to
render them particularly needy. Portraying children as abandoned, however, serves to
pathologise their families and cultures (Burman 1995a). Children are represented as
innocent, which implies someone must be guilty, and with no evidence that they are
supported by family or community, their situation is attributed to the failure of their
people to care for them (Burman 1992).

Portraying children without culture, history or community also serves to avoid addressing
the wider circumstances that contribute to their situation. So too does the exclusive focus

on suffering children (Burman 1992). The American appeal for the 1984 famine in Ethiopia proclaimed 'a hungry child has no politics' (Burman 1992: 243). Children are seen as apolitical and a way of raising emergency funds without admitting to the politics that underlie a disaster situation. Catastrophic images suggest the roots of poverty are in the Third World itself and can be eradicated by foreign aid (Burman 1995a).

Children in aid appeals evoke sympathy by playing on notions of innocence and vulnerability associated with childhood. As these characteristics are also gendered female, it is often girls that are pictured (Burman 1995a). Children generate more support than youth, hence campaigns choose to represent their subjects as children rather than youth, even if they are in their mid-teens: 'to talk of young people as children is to conceptually place them in a (cherished) category to which certain well-defined expectations and entitlements apply' (Hall and Montgomery 2000: 13). Thus young prostitutes in Thailand, represented as victims of child abuse, are viewed with sympathy in Britain, while their 16-year-old London peers, represented as youth, are viewed as vectors of disease and prosecuted as criminals (Hall and Montgomery 2000).

Images of children also encapsulate passivity and helplessness. There is 'an inverse relationship between images of dignity and self-reliance and likelihood to give' (Burman 1992: 249): passive suffering raises more funds than evidence of active struggle. Representing children as passive victims is unlikely to help them cope with trauma (Burman 1992). Using children in this way also serves to cast the Third World in general in the role of recipient of help, as the infantilised beneficiary of Western paternalism. The 1993 ActionAid appeal slogan: 'Please sponsor a child. Tomorrow may be too late' represents the potential donor as an immensely powerful actor (Burman 1992). Similarly, 'Plan International's call to "Change the World One Child at a Time" invites the belief that political and environmental crises can be resolved through helping single children' (Burman 1992: 247).

Aid organisations increasingly recognise the tension between fundraising and the need for public education (Burman 1992). Save the Children (1995) proffers the following guidelines:

1 Preserve people's dignity.
2 Use images that accurately convey the diversity of Save the Children's work – including long-term and self-help.
3 Images should be accurate, not stereotypes.
4 Portray people as active partners in development not passive aid recipients.
5 Portray disabled people as integral to communities, and avoid wheelchairs as they reinforce a stereotyped image of disability.
6 Don't exclude particular ethnic groups, women or disabled people.
7 Avoid patronising, mawkish, sentimental or demeaning text.
8 Unless people want to remain anonymous, try to identify and quote the people who appear in the photograph.
9 Don't crop or edit the picture in a way that distorts the situation, such as removing adults to leave children appearing isolated.
10 Use images in context – don't use photographs from one event or country to illustrate a point about another.

(see Chapter 7). In contrast to earlier declarations of children's rights, it offered children both protection and enabling rights. Although controversial in this respect, it has been the most rapidly ratified international convention in history, due perhaps to the symbolic value attached to children – no government wants to admit that its children are badly treated. Although 'some countries have entered reservations to this Convention that render their ratification of it almost meaningless' (Kuper 2000: 45), it has had a tremendous influence on government and NGO policy.

The year the CRC came into force (1989) was the year set for the World Summit for Children (WSC). This was attended by 71 heads of state who adopted a Declaration and Plan of Action. Heads of state and governments promised to deliver a set of goals that, compared with the CRC, were uncontroversial, measurable and relatively easily achievable (Temba and de Waal 2002). Few countries have, however, achieved the goals and, overall, progress against these modest measures has been limited (Table 1.2).

In 2000, a new set of targets – the UN Millennium Development Goals – were announced. These are not confined to the interests of children, although they include pledges to achieve universal primary education and reduce child mortality by 2015. In March 2001, the Global Movement for Children (a coalition of international NGOs and UNICEF) launched a worldwide campaign 'Say Yes for Children'. Ninety-five-million people pledged their support, most of them children. Signatories were also asked to identify action priorities. The three highest priority issues were education, discrimination and poverty (UNICEF 2003d). In May 2002, the UN Special Session on Children reviewed the previous decade's progress against the goals of the WSC and pledged new goals for the coming decade. This was attended by over 400 children from more than 150 countries (UNICEF 2003d). The Global Plan of Action for Children that developed from the Special Session identified four priority areas of action: promoting healthy lives; providing quality education; protecting against abuse, exploitation and violence; and combating HIV/AIDS.

Youth

Youth have had much less prominence internationally than children. This may relate partly to the popular image of children as innocent, apolitical and outside of economics – an image that cannot be

Table 1.2 *Major goals for child survival, development and protection set by the World Summit for Children, 1990*

Goal	Result
(a) Between 1990 and the year 2000, reduction of infant and under-5 child mortality rate by one-third, or to 50 and 70 per 1,000 live births respectively, whichever is less.	Average global under-five mortality (U5MR) declined by 11% globally, from 93 deaths in the early 1990s to 83 deaths per 1,000 live births in 2000. Over 60 countries achieved the targeted one-third reduction.
(b) Between 1990 and the year 2000, reduction of maternal mortality rate by half.	There is a lack of data. Although skilled care at delivery has increased across all developing regions, in some countries and in sub-Saharan Africa as a whole, where maternal morality is highest, delivery care has not improved significantly.
(c) Between 1990 and the year 2000, reduction of severe and moderate malnutrition among under-5 children by half.	Underweight prevalence declined from 32% to 28% in developing countries over the decade. The most remarkable progress was in East Asia and the Pacific.
(d) Universal access to safe drinking water and to sanitary means of excreta disposal.	Global drinking water coverage rose from 77% to 82%; global sanitation coverage increased from 51% to 61%.
(e) By the year 2000, universal access to basic education and completion of primary education by at least 80% of primary school-age children.	The net primary enrolment/attendance ratio increased from 80% to 82%. The gender gap halved but remains a concern in three regions.
(f) Reduction of the adult illiteracy rate (the appropriate age group to be determined in each country) to at least half its 1990 level with emphasis on female literacy.	Illiteracy decreased from 25% to 20% but the number of illiterate people remained the same, in part due to population growth.
(g) Improved protection of children in especially difficult circumstances.	Not reported.

Source: UNICEF (2001b, n.d.).

sustained in relation to youth, who are undeniably self-willed political beings and undeniably engaged in (or disengaged from) economic life. Nonetheless, the UN has been involved in promoting the interests of youth (or what are seen as interests with respect to youth) over the past four decades. In 1965 the UN General Assembly endorsed the Declaration on the Promotion Among Youth of the Ideals of Peace, Mutual Respect and Understanding Among Peoples. This Declaration exemplifies a view of youth as embodying hope for the future, and the aspirations of society (even the world) as a whole. Youth are seen as idealistic and also impressionable: doing the right thing with youth is to secure the future. At the same time, failing to adopt the right policies with respect to youth jeopardises the future of society. This ambivalent view of youth as both promise and threat continues to pervade UN discourse.

Youth were seen as a major social problem in the Third World in the 1960s, because of rising youth unemployment and because school leavers were moving into the cities in the (unfulfilled) expectation of finding work (Commonwealth Secretariat 1970). As in the West, youth were seen as a potential threat:

> Youth cannot be ignored or kept waiting. Youth is ambitious, and energetic, yet frustrated because its talents are being wasted. When ambitions are thwarted by society which denies youth the opportunity to develop its potential, a serious and explosive situation can arise. Potentially, therefore, this is our most acute social problem, and the greatest source of instability in the nation.
>
> > Opening address by the Hon. Mwai Kibaki, Kenya's Minister of Commerce and Industry (now President), at the Commonwealth Africa Regional Youth Seminar (Commonwealth Secretariat 1970: 4–5)

Proposed solutions focused on providing suitable training, and encouraging/enabling young people to stay in rural areas and to contribute to rural development: to channel their energies in suitable directions through, for instance, Training for Self Reliance (see Chapter 4). There were also efforts to cultivate voluntarism and encourage a degree of participation in planning youth programmes (long before the idea that children should participate was popularised) (Commonwealth Secretariat 1970). To an extent, this reflects the UN's three key youth themes that have persisted in all declarations since 1965: participation, development and peace.

In March 1996, the UN General Assembly adopted the 'World Program of Action for Youth to the Year 2000 and Beyond'. This encompasses commitments related to youth made by signatory governments at various international gatherings, including the youth-specific recommendations in declarations from several international conferences (ICRW 2001). The World Programme of Action incorporates ten priority areas (Box 1.6), with several 'proposals for action' in relation to each. Five further concerns were added later. The emphasis of the Declaration is on young people 'acquiring productive employment and leading self-sufficient

---

Box 1.6

World Programme of Action for Youth to the Year 2000 and Beyond – priority areas

- Education
- Employment
- Hunger and poverty
- Health
- Environment
- Drug abuse
- Juvenile delinquency
- Leisure-time activities
- Girls and young women
- Full and effective participation of youth in the life of society and in decision-making
- *Globalisation*
- *Information and communication technologies*
- *HIV/AIDS*
- *Youth and conflict prevention*
- *Intergenerational relations.*

Source: United Nations (1996, 2002).

---

lives' (United Nations 1996: 6), as well as 'fighting social ills such as drug abuse, juvenile delinquency and other deviant behaviour' (p. 23). However, it also acknowledges that young people are 'confronted by a paradox: to seek to be integrated into an existing order or to serve as a force to transform that order' (p. 4). The World Programme of Action is not legally enforceable like the CRC, but expects commitment from governmental, intergovernmental and non-governmental organisations and institutions at national, regional and international levels (United Nations 1996; see Table 1.3).

There is no 'youth' equivalent of UNICEF, but the Youth Unit in the Division for Social Policy and Development constitutes a very small part of the UN. Every two years the General Assembly of the UN holds a session on youth. There have also been four sessions of the World Youth Forum of the UN System. These are not intergovernmental meetings like the WSC, and have fewer delegates. They have, however, produced guidelines, including the Braga Youth Action Plan and the Dakar Youth Empowerment Strategy.

Table 1.3 *Summary of actions taken by governments to implement the World Programme of Action for Youth to the Year 2000 and Beyond*

Category	Number	Percentage
Countries that have formulated a national youth policy (cross-sectoral)	155	82
Countries that have designated a national youth coordinating mechanism (such as a ministry, department, council or committee)	168	89
Countries that have implemented a national youth programme of action (operational, voluntary service)	116	61

Source: Adapted from United Nations (2001a).

Conclusions

> For their own protection, nurture and enlightenment, children in Western societies are excluded from work and other such responsibilities and confined, largely, to the home and the school, where they experience a prolonged period of social immaturity and dependence. These are the conditions and circumstances that are thought to best favour children's psycho-social well-being and development. Thus, children who do not enjoy such life circumstances are believed to be at risk, their development and adaptation to society undermined.
>
> (Boyden and Gibbs 1997: 22)

Despite the fact that the global model of childhood is clearly based on a Western middle-class ideal, in combination with paradigms from developmental psychology that have long been challenged by academics, it continues to dominate development policy (Boyden 2000). The model has had three principal problematic effects. First, problems affecting children in the Third World are attributed to individual dysfunctioning or pathology (Boyden 1990). Children who do not fit the global model are defined as abnormal (Boyden and Ennew 1997), and the significance of the broad social and economic contexts in which they live is neglected. Second, the global model places responsibility for children's upbringing firmly within the family. Children are not held individually responsible for their own problems, as they are not conceived of as having agency. Rather, it is families that are blamed. Although the notion that children's needs are best served within a family context is not based on empirical research (Boyden 2000), families are expected to fulfil a range of functions, and failure to do so attracts blame. Children are expected to grow up within the home – those found on the streets are considered a problem. Boyden (1990) argues that the main reason for keeping children off the streets is fear of pollution: children are perceived as a threat in need of adult control. Once again, it is the task of families to keep children within the home. Third, because the global model individualises the difficulties children face and places responsibility with families, children are seen as apolitical. Governments and international organisations are able to shift responsibility from themselves onto families, a situation that is exacerbated by the fact that children's NGOs and UNICEF focus most on those issues affecting children that are least politicised and uncontroversial.

Concerns about youth are also inspired by fear. Through a Western lens, delinquency is a natural condition of youth who are not correctly raised. Moral panics around youth inspire a view that, families having failed, society must intervene to preserve itself, both today and for the future. Concern is focused not on young people themselves, but on what might happen if they are not kept in check. Where young people are not viewed in negative terms, the emphasis is on 'youth-as-fun': largely harmless but not meriting serious attention. Such stereotypes detract attention from the positive contributions of young people to social, cultural, economic and, where permitted, political spheres.

Key ideas

- Childhood and youth are understood in very different ways in different societies and at different times.
- A number of contradictory ideas from the past have fed contemporary popular notions of childhood and youth in the West.
- Childhood and youth have been studied in very different ways in different academic disciplines.
- Western popular notions of childhood and dominant academic paradigms have been exported worldwide as a 'global model'.
- According to these global models, children are passive, vulnerable and the responsibility of their families; youth are inherently troublesome and 'at risk'.
- In recent decades there have been numerous international actions designed to improve the situation of children and youth worldwide.

● **Because international actions are based around the global model of childhood, they do not always serve the interests of those who do not conform to the model.**

Discussion questions

- Consider how childhood and youth have changed over several generations of your own family. What (if anything) do you consider to be better today? What (if anything) was better in the past? On what basis do you judge what is 'better'?

- 'It is hard to see how making children disappear from our TV screens or newspapers advances their rights . . . One of the photographs credited with changing Western opinion over the Vietnam war is the most famous picture by Nick Ut of nine-year-old Kim Phuc, running up the road outside the village of Trang Bang, naked and crying. The photograph, taken in 1972, showed the horror of war through the image of a child. Kim Phuc was not asked for her permission, and nor did a social worker give consent' (McIntyre 2002: 51). Should this photograph have been used?

- Think about the terms 'child', 'youth', 'young person', 'teenager' and 'adolescent'. In what contexts are they generally used? What age groups do they usually apply to? What images do they conjure up?

Further resources

Books

Amit-Talai, V. and Wulff, H. (eds) (1995) *Youth cultures: a cross-cultural perspective*, London: Routledge – another edited collection with some useful chapters outlining theoretical approaches to studying children and youth.

James, A., Jenks, C. and Prout, A. (1998) *Theorising childhood*, New York: Teachers College Press – a comprehensive explication of the new social studies of childhood.

Skelton, T. and Valentine, G. (eds) (1998) *Cool places: geographies of youth cultures*, London: Routledge – an edited collection with a useful introduction to the study of youth cultures.

Journals

Relevant articles are published in a variety of journals including *Childhood*.

Websites

Save the Children http://www.savethechildren.org.uk/ contains a history of the organisation.

UNICEF http://www.unicef.org/ contains detailed pages of history as well as some of the key documents relating to recent events such as the World Summit for Children and the UN Special Session.

Youth at the United Nations http://www.un.org/youth/ has plentiful information about young people.

2 'Development', globalisation and poverty as contexts for growing up

Key themes

- Approaches to 'development' and the neglect of children and youth
- Economic globalisation: debt, privatisation and trade liberalisation
- Growing inequality and poverty.

Globalised discourses of childhood and youth have not been the only international influence on the lives of young people over the past half-century. The general development policies pursued by multilateral and bilateral donors, NGOs and Third World governments have been as, if not more, influential. A number of distinct conceptualisations of 'development' have inspired policy interventions. Few of these have given much attention to the interests of children and youth, but they have shaped the contexts in which young people live and grow up. This chapter begins with an examination of the key development paradigms pursued over the past 50 years, in relation to their impacts on children and youth, and then examines the broader contexts of globalisation, growing inequality and poverty that Third World children inhabit.

Approaches to 'development' and the neglect of children and youth

Modernisation theory

'Development', conceived as a process that would ultimately bring Western living standards to the continents of Asia, Africa and Latin America, entered into public discourse only in 1949, when US President Truman made a speech setting out a new duty for the West: to bring 'development' to the 'underdeveloped' parts of the world (Potter *et al.* 1999). This new interest in the fate of the southern continents was rooted in the geopolitical circumstances of the time, as well as in a broader commitment to 'progress' that had imbued Western thinking since the 'Enlightenment' of the eighteenth century.

The 1940s had seen international cooperation (as well as conflict) on an unprecedented scale during the Second World War, and subsequently the establishment of the United Nations and the international financial institutions (IFIs, forerunners of the World Bank, International Monetary Fund and World Trade Organisation). The late 1940s also marked the onset of the main period of European decolonisation: 70 countries in Asia, Africa and Latin America were to gain independence in the following quarter-century. During the Cold War, Western states and institutions competed for the allegiance of these new nations, defining themselves as change-agents, promising 'development' to what became known as the 'Third World'.

The United Nations resolved that the 1960s should be the 'Decade of Development'. There was optimism that the capitalist conditions that had brought economic growth to the West would work everywhere. It was believed that a single formula, applied at the nation-state level, could bring prosperity for all. This undifferentiated model became labelled 'modernisation theory' and was most explicitly formulated by American economist W.W. Rostow (1960). Rostow understood economic development as a series of stages, movement through which was driven by investment in the productive sectors of the economy and by a modernisation of attitudes.

Modernisation theory has little to say about children or youth. Operating at a macro-economic nation-state level, ordinary people have little specific role. Even the end-point of development – the stage of 'high mass-consumption' – was seen as a stage in which all would benefit equally, without any need to distinguish between the impacts on those of different ages. Given such a theoretical context, it is unsurprising that children and youth were largely by-passed by the development strategies adopted in the 1950s and 1960s. Development was understood to be something undertaken by governments (and later by IFIs), not by individuals, in particular not by those individuals in whom little economic or political power was vested, including children and youth.

The one area of modernisation theory that affected young people was education. Rostow (1960: 182) believed 'the education of a generation of modern men' was required before an economy could begin to grow. Mass education became seen as both a requirement for, and an indicator of, modernity (Inkeles and Smith 1974). This view of education was, however, very instrumental: the fact that its recipients were children was largely irrelevant (see Chapter 5).

Modernisation theory underlay the subsequent development of human capital theory: economic growth was understood to require investment in the education, training and health of the workforce. Lack of skills and education among labour forces in Third World countries were believed to hold back productivity. Children were a resource to be invested in (UNICEF 1996): schools would provide them with knowledge and skills to participate productively in the economy, while technologies such as immunisation would improve the health of the future workforce.

Modernisation-inspired policies had impacts on young people beyond schooling and immunisation, however. Investment in the 'modern' sectors of the economy, particularly industrialisation, led to high rates of urbanisation. Urban life for children and youth differed considerably from that which their parents had experienced in rural areas, but was not necessarily an improvement. Most expected formal employment: an expectation that generally remained unfulfilled. Furthermore, there remained large numbers of people, resident in rural environments, for whom the 'modern' life of the city remained a distant dream.

Third World socialism and dependency theories

Some of the nation states achieving independence in the latter half of the twentieth century rejected Western development formulas, choosing instead to pursue policies based on Marxist ideological foundations. Marxist-inspired governments have at various times held power in countries as diverse as China, Chile, Cuba and Ethiopia. The implementation of Marxism differed substantially from nation to nation, but in general land, mineral resources and industries were nationalised, foreign trade and investment brought under state control, and rigid price controls exercised (Kilmister 2000). The intention was to achieve development that would benefit the lower classes and not just the owners of capital. Some countries chose a route based on heavy industrialisation, little different from the Western-sanctioned modernisation pursued elsewhere. Others, notably China, preferred a more agrarian-based course.

Beyond a vociferous condemnation of child labour (Marx 1976/1867), Marx's writings say little specifically about children and youth. The family is, however, a focus of attention among Marxist theorists: the communist manifesto (Marx and Engels 1998/1848) sees the family as a locus of exploitation and calls for its abolition. Functions previously fulfilled on a private basis through the family would be replaced with social provision. These ideas have had their most extreme expression in Cambodia/Kampuchea under the Khmer Rouge where, in some areas, families were forcibly broken up and children taken away to be raised by the Angkar (the ruling organisation).

Few Marxist regimes have gone so far as to abolish the family, but most have seen social investment in children as desirable (Box 2.1). Children are seen much less as the property of individual families than is usually the case under capitalist regimes, and are considered legitimate subjects of government intervention. As with modernisation theory, children are seen as the nation's future, and their health and education is considered a worthwhile focus of government energies. Education is valued not only for its role in improving economic productivity, but also as a tool to encourage young people to take on particular traits and accept the state ideology. Hence, Marxist governments have generally invested heavily in ensuring that education is extended to as many young people as possible. In many cases, governments have also established other means whereby young people can be persuaded to commit themselves to the ideological cause, including youth leagues and brigades.

The 1970s saw the arrival of a new theoretical basis for rejecting Western capitalism. Prescribed investment was not bringing the

Box 2.1

Social policy in Cuba

Following the revolution in 1959, Cuba's government had a vision of *sembrando el futuro* (sowing seeds for the future), and invested in primary health care with an emphasis on preventive care through community-based clinics (Whiteford 1998). There were four national health goals: preventive medicine; improved sanitation; raising nutritional levels among disadvantaged groups; and health education. Medical care was made available free of charge; the number of health workers was increased; medical facilities in the rural areas were improved; water was chemically treated; malaria eradicated; food rationed; and early intervention promoted in relation to problem pregnancies and diarrhoeal diseases (Whiteford 1998). The emphasis was on both access and equity. By 1991, Cuba had 38,000 doctors – one for every 534 people (compared with one for every 714 people in the UK). Each was assigned 120 families and provided with housing within the community. In 1992 well-baby home visits were made daily for the first week of an infant's life, weekly for the first month and then twice monthly for the first year, enabling health personnel to get to know the whole family and neighbourhood, and become aware of any other health problems. Cuba has also focused on eliminating illiteracy and has arguably the best education provision in Latin America (McLaren and Pinkney-Pastrana 2001).

Cuba's government continued to increase its allocation of resources to education and health, despite global economic recession (Muniz *et al.* 1984). Even with the loss of support from the former USSR from 1989, and under the US embargo (extended in 1992 by the Helms Burton Act which prohibits any company trading in the USA from trade with Cuba), Cuba maintained steady improvements in health status. Public hospitals suffered from resource shortages, fewer non-essential procedures were performed, and the number of ambulances declined, but community-based clinics were not badly affected. Although diarrhoeal diseases and infectious diseases increased slightly in the early 1990s, overall infant mortality rates continued to fall (Table 2.1).

Plate 2.1 *Crossed pencils symbol of Cuba's 'literacy army'.*

Source: Author.

Table 2.1 *Child welfare in Cuba, compared with the USA, Latin America and the Caribbean*

	Cuba	USA	Latin America and Caribbean
Child mortality 1960	54	30	153
Child mortality 2001	9	8	34
Infant mortality 1960	39	26	102
Infant mortality 2001	7	7	28
% infants with low birthweight	6	8	9
% under-5s underweight	4	1	8
% primary school entrants reaching grade 5	95	99	76
Health as % of government spending	23	21	6

Source: Data from UNICEF (2003d).

rewards that modernisation theory had promised. Indeed, far from 'underdevelopment' being eradicated by the end of the 1960s, inequalities between the West and its Third World satellites were growing. Even where economies had grown, the poor saw few improvements in their lives. While some blamed nation states for failing to deploy development policies in the prescribed ways, an increasing body of scholars, particularly in Latin America, felt the blame for 'underdevelopment' should be cast elsewhere.

Inspired in part by Cuba's relative success, despite the US trade embargo imposed after the 1959 revolution, a number of scholars concluded that the operation of the global economy, far from helping countries to 'develop', was the reason for their state of 'underdevelopment'. Andre Gunder Frank (1967) expounded the idea that 'underdevelopment' was not simply the condition of societies which had not yet undergone 'development', but was rather the obverse side of the active process whereby Western nations became 'developed'. This 'dependency theory' viewed the global economic system as an engine by which the West creamed off surplus value generated through the extraction and trade of raw materials from the South. This structuralist model accorded little importance to human agency, people being cast largely in the role of victims or beneficiaries of an impersonal system.

Unlike modernisation theory, dependency theorists saw development as a process with differential impacts. The differences it recognised were, however, largely geographical. A relation of dependency was shown to exist between cores and their peripheries, the cores creaming off surplus profit from the peripheries. This process operated both at the global level, between the West and the Third World, and at more local levels, including between cities and their hinterlands. As another macroeconomic model, little attention was given to differences within societies.

The policies inspired by dependency theory differed little in their impacts on young people from those inspired by modernisation theory. While Frank (1967) suggested the solution to Third World dependency – complete isolation from the global capitalist economy – could be achieved only through revolution, dependency theory's practical impact was less dramatic. A number of countries opted for import-substitution industrialisation – by establishing industries producing for local consumption, they would be less dependent on overseas trade, and thereby less vulnerable to the vagaries of the international marketplace. Many, such as India, did so as part of a policy of 'non-alignment', seeking to avoid dependence on either the West or the USSR. Import-substitution industrialisation affected young people in much the same way as export-oriented industrialisation.

Neo-liberal development and structural adjustment

Western banks and donors loaned heavily to most Third World countries on the basis of the promises of modernisation theory. This investment was partly fuelled by Cold War fears. Also, following the Organisation of Petroleum Exporting Countries (OPEC) oil price hike of 1973, large amounts of 'petrodollars', earned by oil-producing countries and invested in international banks, were loaned out at low interest rates. The borrowing failed, however, to generate the expected levels of economic growth. By the late 1970s, not only was growth stagnating, but countries were highly indebted. The introduction of monetarist policies in the UK and USA, particularly following the 1979 OPEC oil price hike, brought high interest rates. These were crippling to Third World countries, simultaneously faced with shrinking commodity markets. The threat of default on debt (led by Brazil in 1982) provoked a new policy on the part of the IFIs: the International Monetary Fund (IMF) was willing to reschedule national debts, but only on condition that strict structural adjustment policies (SAPs) were put in place. In stark contrast to the principles underlying import-substitution industrialisation, countries were encouraged to open their economies to global markets. Governments were persuaded to implement a range of SAPs that had direct impacts on children. These included cuts in public spending; the removal of price controls and subsidies, particularly on food; the privatisation of public enterprises, which led to unemployment among public sector workers; and wage restraint (Figure 2.1).

Cuts in public spending had the most immediate impact on children. In most Third World countries, the three main areas of public

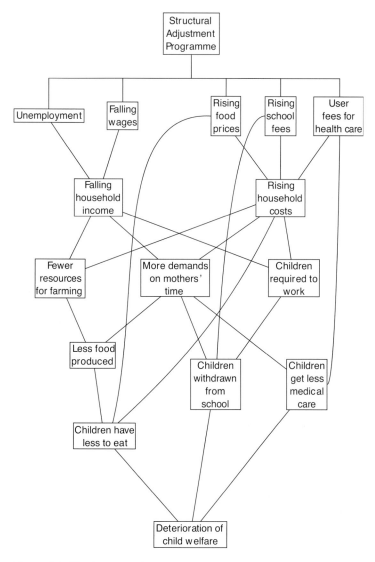

Figure 2.1 *The impacts of structural adjustment policies on children.*
Source: Adapted from Messkoub (1992).

spending are education, health and the military. While there was little encouragement from the West to cut military expenditure (this would have damaged export markets for Western defence industries), health and education spending were targets for reform. Reduced subsidies, combined with increased costs, undermined the quality of social provision. The Dominican Republic, for example, negotiated IMF loans in 1983 and 1985 to compensate for a 70 per cent fall in income from exports between 1981 and 1985, resulting from global economic recession (Whiteford 1998). The conditions attached to these loans required price increases on basic items of over 80 per cent, and a 300 per cent increase in tariffs on imports. Energy costs dramatically increased. Electrical blackouts became common and affected hospitals. Overworked generators frequently broke down, so most of the time there were no lights and operations had to be conducted by light entering open windows. There was also often no running water – hence no flushing toilets and no possibility of washing or cleaning equipment. Over a decade from the late 1970s, levels of severe malnutrition among children in the Dominican Republic doubled, maternal mortality doubled and infant mortality almost doubled (Whiteford 1998).

Many countries required to implement SAPs were encouraged to charge for services which had previously been free (euphemistically termed 'cost sharing'), on the basis that passing costs onto users would enhance efficiency and raise valuable funds to improve services. Government subsidies to schools were frequently reduced and, to compensate for this, primary school fees were introduced and secondary fees increased. Other school-related costs (uniforms, books, exam fees) also rose. In Tanzania, where primary education is compulsory for children over seven, families who cannot afford to pay fees are technically breaking the law (Marcus 2001). Similarly, clinics and hospitals charged patients for visits and for medicines. These changes in public services hit children harder than adult members of society (Messkoub 1992). Children are the main users of education, but also make greater demands on health services than adults. In sub-Saharan African countries, primary school enrolment declined by about 6 per cent between 1980 and 1985 (Messkoub 1992). Families were forced to choose which of their children they could afford to send to school: girls were often withdrawn first. Infant mortality rates increased in some countries for the first time on record (Narman 1995) and under-five mortality rates improved little (Messkoub 1992).

The introduction of user fees for public services reflected not only a particular economic philosophy on the part of the IFIs, but also a distinct conceptualisation of the role of the family. Particularly in former socialist countries, responsibility for children's welfare reverted from the state to the family, with families providing for all the needs of their offspring. In the 1980s and 1990s childhood was being privatised.

The privatisation of children's welfare meant that children's lives increasingly depended upon the survival strategies of people in their households (Messkoub 1992). Thus children were affected not only by cuts in services that directly impacted their lives, but also the 'scissor effect' of increased unemployment, falling wages and rising food prices brought about through structural adjustment (Messkoub 1992). Many poor families found themselves in situations where they were increasingly unable to secure adequate food, with disproportionate impacts on children who are more vulnerable physiologically than adults to inadequate diets.

Another consequence of declining household resources was the need to find new sources of income. Not only is falling school enrolment an outcome of lack of income to pay school fees: children also drop out of school to contribute to household survival. In some cases children are sent out to work; in other cases they are required to perform household tasks (such as care for younger siblings) that were previously the responsibility of other household members who themselves joined the workforce.

Youth, too, were adversely affected by SAPs. Many were withdrawn early from school and at the same time the availability of jobs declined. The labour market in Kenya is typical in its inability to keep pace with the output from formal schooling. Between 1982 and 1990 the number of waged employees in Kenya increased by 280,000 to 1.4 million – an increase of 25 per cent. Meanwhile, an estimated 2.5 million left education to compete for employment. Of those in formal employment in 1990, 40 per cent worked in the public sector, which was to undergo drastic reduction as a result of structural adjustment. Concurrently, manufacturing industries were supposed to be growing more efficient, and less labour intensive. Hence only 25 per cent of all new jobs between 1990 and 2000 were expected to be in the formal sector (Narman 1995).

NGOs and people-oriented development

The neo-liberalism that inspired the IFIs to demand structural adjustment as a condition for lending was also associated with a deep scepticism about the role of the state as an agent of development. This related in part to the perception that Third World governments often lacked democratic mandates and were susceptible to corruption. Through the 1980s and 1990s both multilateral and bilateral donors elected to channel aid increasingly through NGOs in preference to governments (Box 2.2). While NGOs are by no means homogeneous, many are characterised by a different philosophical approach to development, concerning themselves more with ordinary people than with macroeconomics.

NGOs take many forms. Some have long histories, rooted in philanthropic organisations of the early twentieth century, or in religious and missionary establishments with even longer genealogies. Others have been established more recently, as development funds have become available. International NGOs (INGOs) generally have their bases in Western countries, and may not operate 'on the ground' in Third World countries at all, preferring to channel funds through smaller local-level organisations (Lewis 1998). National and local NGOs in Third World countries are often set up with the hope of securing funds from INGOs or donors in order to achieve certain outcomes. Such NGOs commonly give attention to the welfare of particular groups within society, including women, ethnic minorities and, increasingly, children and youth.

The discourse of NGOs has focused on 'people-oriented' development. In the 1970s policy-makers grew sceptical of technological solutions and began to give attention to the 'felt needs' of the poor (Scheper-Hughes and Sargent 1998). Despite high rates of economic growth in some countries, there had been little improvement and in some cases deterioration in the life chances of the world's poor – particularly children (Scheper-Hughes and Sargent 1998). Development had not 'trickled down' to children, as anticipated (Johnson 1996). In the 1980s, NGOs operating in Third World countries became aware of the negative impacts of SAPs, particularly on the more vulnerable members of society. NGOs began to emphasise a 'basic-needs' approach to development, in which meeting the welfare needs of poor people would be the key priority for investment (Potter *et al.* 1999). Those whose needs were recognised as particularly acute included women and children – often

---

Box 2.2

The state versus NGOs in Latin American child welfare

Child welfare in Latin America was provided through charity and philanthropy from the early colonial era until the early twentieth century. Formal state provision began in the 1930s, when several Latin American countries established legal frameworks known as 'children's codes' to address both abandoned and neglected children and those committing offences. Most children were placed in residential care facilities. Because the welfare state in most Latin American countries was weak, particularly in relation to the scale of poverty, such children came under the remit of the law enforcement agencies. Although the intention was to rehabilitate, most children dealt with through the courts were effectively imprisoned in institutions, whether or not they had themselves infringed the law. The approach did nothing to tackle the underlying reasons why children and their families came into conflict with the law. Thus, while the middle classes were permitted to raise their children privately, with support from the state, those living in poverty were exposed to intrusive state intervention.

In the 1970s and 1980s there was a dramatic surge in the number of NGOs involved in child welfare issues. This partly reflected the prevalence of authoritarian regimes in the 1970s that neglected the poorer sections of society and excluded numerous professionals, particularly social scientists, from universities and government agencies. Also at this time, much bilateral and multilateral aid was directed into NGOs rather than government. NGOs generally took a more preventative approach, emphasising family and community participation. They were also able to respond to situations more quickly than state bureaucracies, and sought to avoid removing children from their environments. However, the focus was largely on preschool children and their mothers, older children remaining the preserve of the courts.

Source: Pilotti (1999).

---

bracketed together under the assumption that women's main role was the care of children and children were the responsibility of their mothers. Children were particularly attractive to NGOs, as the fact that they are perceived as apolitical finds favour with both funders and host governments.

A needs-based approach to development for the first time focused the thoughts of development practitioners on the specific situations and needs of children. It was recognised that children have interests that are distinct from adults (Table 2.2). However, this approach to development was not unproblematic. Women and children had previously been neglected on account of their (assumed) invisibility

Table 2.2 Key issues identified by children and other members of the community at Kyakatebe, Masaka District, Uganda

Issues of concern	Team identifying the issues	
	Children	Adults
Child labour	✓	–
Transport to school	✓	–
Inadequate food	✓	–
Drunken teachers	✓	–
HIV/AIDS	–	✓
Lack of fuelwood	–	✓
Inadequate health facilities	–	✓
Lack of school fees	✓	✓
High level of school opt-out	✓	✓
Orphans	✓	✓

Source: Adapted from Johnson and Ivan-Smith (1998).

in the productive economy: they were now cast in the role of passive recipients of aid. NGOs later began to see women as not simply recipients, but resources for development, involving them in 'participatory' development projects (Nelson and Wright 1995). Many such projects were exploitative, taking advantage of women's unpaid labour to perform work previously the responsibility of the state. They did, nonetheless, accord women the status of actors. Children, by contrast, remained objects to be provided for by experts. Furthermore, women's and children's interests do not always coincide. Household income generation schemes may, for instance, mean the withdrawal of children from school, either because child labour is itself valuable in the enterprise, or because mothers' participation in such schemes means that child care responsibilities are passed on to girls. Equally, measures aimed at improving children's lives sometimes have costs for women (Burman 1995a).

From the mid-1980s onwards, women began to argue that questions of power in the relations between men and women needed to be addressed, if women's lives were to improve (Molyneux 1985). This gender analysis fed into a broader 'empowerment' approach to development, which began to inspire those working with children and youth. Over the past decade, NGOs have begun to advocate 'rights-based' approaches to development. The application of a rights discourse to policies relating to children and youth has drawn particularly on the United Nations Convention on the Rights of the

Child (1989) discussed in Chapter 1. This has contributed to a growing recognition that children are not merely passive 'beneficiaries' of development policies and projects, but key actors in development.

Economic globalisation: debt, privatisation and trade liberalisation

NGO-led development is, on the whole, reactive. NGOs generally have only sufficient resources to engage in small-scale activities within a world in which change is being fuelled by processes that are well beyond their control. Even budgetary subsidies (see Box 2.3), although significant to the governments of small nation states, have little impact compared with wider global economic forces. It is necessary to recognise that, as a backdrop to NGO and government programmes working with young people, economic globalisation is having impacts which are more far-reaching than any of the deliberate actions of the development industry.

As conditions on debt rescheduling (Box 2.3) and aid, Third World governments have been encouraged to open their economies to free trade on global markets. The idea is that expanding exports in areas in which they have 'comparative advantage' (usually primary commodities – minerals, tea, coffee, etc., or industries that take advantage of cheap relatively unskilled labour), will allow their economies to grow more rapidly. The impacts of globalisation affect different types of households differently. Foreign direct investment (generally meaning multinational investment in processing plants) brings jobs in export processing industries, but most of the jobs go to childless young women: youth find employment but there are few direct benefits for children (Hilary 2001). Sometimes the export sector employs children, usually in home-based manufacture of components or in agricultural work, but this is of questionable value to children (see Chapter 6). Disinvestment is relatively easy, hence there is no longer a need for businesses to nurture the next generation of employees locally. This spatial separation of the employment and social reproduction of labour has consequences for children and its impacts are under-researched (Aitken 2001).

Prices on world markets are often highly volatile, and Third World producers are vulnerable. The recent rapid expansion of coffee production in Vietnam is undermining prices and livelihoods in Latin

Box 2.3

The impacts of Third World debt on children

Despite structural adjustment, Third World debt continued to deepen in the 1990s. Many of the worst affected countries are also among the poorest. Of 41 countries classified as Heavily Indebted Poor Countries (HIPCs), 30 are categorised as LDCs (Least Developed Countries). Yet these impoverished countries are required to spend a very substantial part of their budgets on debt servicing (interest and repayment) (Figure 2.2). On average, LDC budgets are allocated 5 per cent to health, 13 per cent to education, 14 per cent to defence and 20–30 per cent to debt servicing (UNICEF 2001a). Unable to devote much spending to health, it is unsurprising that the debtor countries have poor records for child survival (Table 2.3). Of the 30 countries with the worst child death rates, 25 are HIPCs.

In the mid-1980s, the President of Tanzania, Julius Nyerere, asked: 'Should we really let our children starve so that we can pay our debts?' (cited in UNICEF 1999: 2).

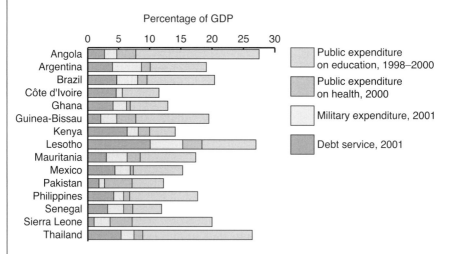

Figure 2.2 *Public spending on external debt and social services for selected countries.*
Source: Data from UNDP (2003).

Table 2.3 *Human development indicators for HIPCs*

	Industrialised countries	Developing countries	HIPCs
Under-five mortality per 1000 1997	7	96	155
% under-fives underweight 1990–1997	–	31	43
Life expectancy at birth 1997	78	63	53

Source: Based on UNICEF (1999).

The international community has, however, been slow to address the problems of indebtedness. Debt rescheduling in the 1980s and 1990s gave a measure of debt relief, but was insufficient to avert continued accumulation of debt arrears and it also excluded multilateral creditors (notably the IMF and World Bank). By the mid-1990s, 40 per cent of bilateral aid was being used to pay off multilateral debt, and the bilateral donors were eager for change (UNICEF 1999). The HIPC initiative, launched in 1996, was intended to make debt sustainable through achieving a balance between debt and export earnings. It also covered multilateral debt and allowed for debt reductions of 80 per cent or more in some cases. It was, however, a very slow two-stage process and by mid-1999 only four countries had received any debt relief. The G8 summit at Cologne in 1999 agreed an Enhanced HIPC Initiative, intended to tackle poverty more directly, but for the first 22 countries debt service payments have fallen by only 27 per cent (Jubilee Debt Campaign 2002).

HIPCs receive more in new aid than they repay on debt, which has prompted some to suggest debt servicing is irrelevant to financing the social sector. Yet aid is highly erratic, which undermines budget planning, and is often spent on off-budget activities, particularly capital expenditure on projects. In Tanzania only one-third to one-half of donor money passes through the national budget (UNICEF 1999). In acknowledgement of these difficulties, some donors have recently begun to switch from funding NGO projects to budgetary support for governments.

America (Hilary *et al.* 2002). During the Asian economic crisis of the late 1990s, international finance rapidly withdrew from the Philippines, Thailand, South Korea and Indonesia, banking went into crisis and unemployment rose dramatically. Thousands of children dropped out of school and, particularly in Indonesia, child malnutrition increased. Malaysia, which had stricter controls on finance flowing in and out of the country, suffered much less social disruption (Marcus 2001). Although such financial crises are temporary, their negative impacts on children are disproportionately severe and long lasting. Children may be removed from school, and be unable to return; lack of capacity to purchase food can result in malnutrition, causing permanent brain damage; some children enter hazardous employment, including prostitution, which can have lasting effects (Hilary *et al.* 2002).

Foreign private sector investment and privatisation are also affecting the provision of basic services in Third World countries (Hilary *et al.* 2002). Under the World Trade Organisation's (WTO) General Agreement on Trade in Services (GATS – see Box 2.4) governments must allow foreign companies to compete to provide services without applying 'unnecessary' quality standards. GATS undermines

---

Box 2.4

WTO agreements

The WTO is opposed to the use of performance and entry requirements on foreign investors.

- *TRIMs Agreement (Trade-Related Investment Measures)* prohibits governments from providing subsidies to firms because they use domestic rather than imported goods.
- *SCM (Agreement on Subsidies and Countervailing Measures)* prohibits governments from providing subsidies to firms because they use domestic rather than imported goods.
- *Agreement on Agriculture* requires governments to liberalise agricultural markets. This threatens food security by allowing countries to become dependent upon unreliable foreign imports.
- *TRIPs (Trade-Related Intellectual Property Rights)* obliges countries to protect intellectual property. This restricts access to patented commodities such as medicines.
- *GATS (General Agreement on Trade in Services)* bans governments from imposing conditions on foreign investment in services.

Sources: Hilary (2001); Hilary *et al.* (2002).

---

countries' abilities to regulate their own health services, or even to determine their own public health priorities (Hilary 2001). A popular arena for foreign investment has been the water industry. The French company, Vivendi, for instance, gained a 30-year concession to supply water to Tucuman province, Argentina, and immediately the price to customers rose by 100 per cent (Hilary *et al.* 2002). Similarly, in the Philippine capital Manila, private water providers, jointly owned by French and US companies in partnership with domestic owners, persuaded the public water regulator to allow a 50 per cent increase in water rates in August 2001, despite their failure to provide the contracted 24-hour service. Poor families, unable to pay the higher rates, are forced to collect water from untreated sources, risking exposure of children to water-borne diseases (Hilary *et al.* 2002). 'Lock-in' clauses make liberalisation under GATS irreversible (Hilary 2001).

In view of the shortage of donor and government funding, foreign investment in health services might seem desirable. Private sector health care and health insurance companies from the USA and Europe are already expanding into Latin America. This has, however, exacerbated problems of equity, quality and capacity. Whole communities are excluded as they are unable to pay. New service providers focus on the most profitable market segments such as urban areas where provision costs are lower and incomes higher (known as 'cream skimming'). Government support to other areas may end, resulting in a rise in prices or decline in availability for poor areas (Hilary *et al.* 2002). Moreover, commercial considerations can result in a downward pressure on quality and key medical personnel are drawn away from the government services to serve only the healthiest and wealthiest consumers (Hilary 2001).

In many cases, economic liberalisation has had negative impacts on children's health (Box 2.5). In China, the replacement of the commune health system with user fees and private health care has witnessed increased incidence of TB and schistosomiasis, as well as declining child immunisation rates and increases in stunting and child mortality (Hilary *et al.* 2002). A survey in Cambodia found 45 per cent of rural families who had lost their land had done so because of debts relating to medical expenses – a situation exacerbated where health care is provided on a profit-making basis (Hilary *et al.* 2002).

The World Bank admits that the net gains to sub-Saharan Africa and South Asia from trade liberalisation will be minimal (World Bank 2002), yet this approach persists and is advocated by many bilateral donors as well as the multilaterals. Even NEPAD (the New Partnership for Africa's Development) is based on a premise that trade liberalisation is the answer to Africa's economic problems (Save the Children 2002a).

Growing inequality and poverty

There has been a marked increase in economic disparities, both across and within countries in recent years (Figure 2.3). Between 1960 and 1993 the gap between the average per capita income in industrialised and developing countries tripled. The poorest countries dropped even further behind (Fussell and Greene 2002). The UN defines 48 countries as Least Developed Countries (LDCs) on the basis of: income (Gross Domestic Product (GDP) below $900 pc); quality of life (life expectancy, calorie intake, primary and secondary enrolment rates and adult literacy); and economic diversification (share of manufacturing in GDP, share of labour force in industry, per capita commercial energy consumption, merchandise export

---

Box 2.5

Economic reform in Central Asia: impacts on children

Following the collapse of the USSR in the late 1980s, the Central Asian region lost its export markets, and rapidly implemented economic reforms. In the 1990s, exchange rate instability led to falling output across Central Asia, accompanied by falling wages, rising unemployment, growing inequality and reduced public spending on social services. Income poverty increased over the 1990s from 0 to 36 per cent in Mongolia; maternal mortality rates doubled from 12 to 24 per 10,000 from 1991 to 1993, while public health expenditure fell from 5.5 to 4.0 per cent of GDP. Spending on education fell to about a quarter or a third of pre-independence levels across Central Asia, resulting in falling school enrolment. Gini coefficients (a measure of economic inequality) have increased across the region (Wilkinson 2000b).

It is argued that children have suffered most in economic transition (Falkingham 2000). Economic dislocation diminished material welfare at the household level (reducing the capacity to pay for health and education and increasing the need for social services), while reducing public provision of education, health and culture. Research in Kyrgyzstan revealed that child poverty increased significantly between 1993 and 1996, and the rate of poverty among children was higher than among the population as a whole, the worst affected being younger children and those from rural areas. Infant mortality rates increased in all the former Soviet Central Asian countries in the early 1990s, but remain low compared to other low- and middle-income countries. Currently immunisation rates remain almost universal and primary school enrolment rates are high. Levels of stunting and wasting, however, are now akin to those in parts of sub-Saharan Africa.

Source: Falkingham (2000).

---

concentration) (UNICEF 2001a). Over the 1990s, the income gap between LDCs and other Third World countries widened from 4.5 times to 5.2 times less (UNICEF 2001a). The 30 African LDCs fared particularly badly, per capita Gross National Product (GNP) falling in both the 1980s and 1990s (UNICEF 2001a). Aid to LDCs also fell during the 1990s, which is significant as aid accounts for a large proportion of their external resources (six to eight times as much as foreign direct investment) (UNICEF 2001a).

The growing income disparities are reflected in disparities in children's health and survival. Although LDCs are home to only

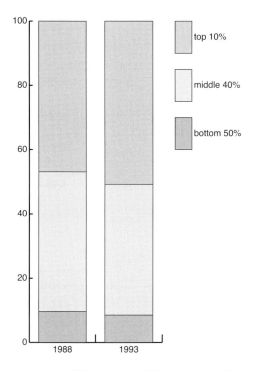

Figure 2.3 *Widening global income disparities.*
Source: Data from Milanovic (1999).

10 per cent of world population, because birth rates are high, they account for about 20 per cent of births and a disproportionate number of children (UNICEF 2001a). The under-five mortality rate in LDCs fell from 182 to 162/1,000 over the 1990s – a proportionately smaller decline than for other Third World countries (UNICEF 2001a). In sub-Saharan Africa, both infant mortality and under-five mortality increased in the late 1990s (Wilkinson 2000b). Underweight prevalence among preschool children is increasing by 0.56 per cent a year in eastern Africa and 0.32 per cent in western Africa (Wilkinson 2000b).

There are also growing disparities within countries. Even where the numbers living in poverty are diminishing, *severe* poverty is commonly affecting more of the population (see Table 2.4). In Brazil, the under-five mortality rate is 113.3/1,000 among the poorest 20 per cent of the population, but only 18.7/1,000 among the richest 20 per cent (Wagstaff and Watanabe 2000). In many countries, children are over-represented among the poor (Figure 2.4): in Vietnam, 63 per cent of children live in poverty,

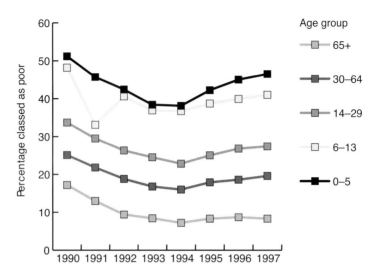

Figure 2.4 *Percentage of each age group classed as poor, Uruguay, 1990–1997.*

Source: Based on UNICEF (2003b).

Table 2.4 *Change in poverty rates (%)*

	Poverty	Severe poverty
Tanzania 1983–1991	−9	+3
Kenya 1981–1991	−3	+4
Nigeria 1985–1992	−14	+11

Source: Based on UNICEF (2000, after Demery and Squire 1996).

compared with only 50 per cent of the population as a whole (Wilkinson 2000b).

Growing inequity is also manifested in a growing social distance between rich and poor. Because of urbanisation, poor children are very visible in Latin America (Pilotti 1999). Yet while downtown areas of Latin American cities used to be cultural, commercial and administrative centres where people from all classes interacted, today, middle- and upper-class children no longer use public facilities like schools and transport, and they shop and socialise in malls, hence they seldom interact with the children of the poor (Pilotti 1999).

Measuring child and youth poverty

Child poverty has become a key concern of NGOs. A typical view is expressed by UNICEF: 'Children are often hardest hit by poverty: it causes lifelong damage to their minds and bodies. They are therefore likely to pass poverty on to their children, perpetuating the poverty cycle' (UNICEF 2000: 1). Although enumeration of poverty is extremely difficult, it is estimated that of 1.06 billion 15–24-year-olds globally, between 38 and 110 million live in extreme poverty, 238 million on less than $1 a day and 462 million on less than $2 a day (United Nations 2002: 5). Poverty is not just about lack of income, however, but also basic needs including health, education, nutrition and shelter, and arguably security and 'empowerment', which can include political participation (Box 2.6). UNICEF advocates the use of the under-five mortality rate as the main indicator of human development, as it is influenced by a wide range of determinants and is not subject to excessive influence from a high-income minority (White et al. 2002). Although correlated with income within and across countries, the correlation is imperfect: rising income will not completely eradicate deprivation (White et al. 2002). UNICEF is also developing a composite 'Child Risk Measure' that uses under-five mortality rates, the prevalence of children born underweight, the percentage of non-attendance at primary school, security ratings and adult HIV prevalence. According to this measure, children are at greatest risk of poverty in Angola, Sierra Leone, Afghanistan and Somalia (Wilkinson 2000b).

> If poverty is defined solely in terms of income, then economic growth will appear to be the best poverty-reduction policy. But as soon as the policy objective is broadened to include, say, health and education, then social policy will assume a more important role.
>
> (White et al. 2002: 4)

In order to understand how children are affected by poverty it is necessary to look beyond broad health and education data. Save the Children argue for more use of non-traditional indicators of poverty (Wilkinson 2000b). Research in India, for instance, revealed that even where income had fallen, poor people felt their welfare had improved in terms of measures they regarded as important, such as wearing shoes and having separate accommodation for people and livestock (White et al. 2002). Save the Children also suggest children's own views and perspectives on poverty need to inform the

Box 2.6

The inadequacy of income as a measure of poverty

- *China* – between 1978 and 1995 the proportion of the population living in income poverty fell from 30% to 10%. However, child mortality stopped falling after 1985, and may have since risen, and stunting is also reported to have risen since 1985.

- *Costa Rica* – per capita income is only about 10% of that of industrialised countries, but life expectancy is about the same. Investment in basic services enabled a halving of adult illiteracy and reduction of child malnutrition by two-thirds, and infant mortality by three-quarters between 1970 and 1990. The poverty headcount index, however, remains unchanged at about a quarter of households.

- *Indonesia* – income poverty rates were reduced from 60% of the population in 1960 to 10% in the mid-1990s. Nonetheless, a third of under-fives were underweight in 1995, a higher proportion than the sub-Saharan African average. Only a fifth of the children of the poorest quintile of the population completed nine years of education – fewer than in Zimbabwe.

Source: UNICEF (2000).

choice of indicators (Wilkinson 2000b). When children are asked about poverty, their concerns generally fall into four categories: survival (availability of food, water and shelter); material well-being (access to money and wealth, schools and work); social support (family, friends, neighbourhood, boredom); personhood and sense of self (nationality and status, mental health, hopes and aspirations, participation, feelings of vulnerability and openness to exploitation) (Wilkinson 2000b).

Much less attention has been paid to poverty among youth. Despite the numbers of young people living in poverty, they are seldom a high priority for policy-makers, as they are perceived to be resilient and mobile. It is not recognised that young people have a high potential to respond positively to employment and income-generation programmes, and could benefit others in this way (United Nations 2002).

Conclusions

Governments and development NGOs have tacitly acknowledged children's 'difference' through the vast finances devoted to providing schooling, as well as, for example, childhood immunisation programmes, yet policy approaches to young people's needs have at best tended to treat children and youth as 'objects' of development (Johnson and Ivan-Smith 1998). Often, children have represented little more than indicators of societal well-being (through such measures as infant and child mortality rates). Similarly, concern with youth has centred on their employability, with regard to their contribution to economic growth. The emphasis has been less on young people themselves than on what their condition says about, or what they contribute to, wider society. Since the 1980s, growing global economic inequalities have left many children in situations of worsening poverty. Yet, paradoxically, the neo-liberal policies that have fed poverty have also led to funds being directed away from government-led macroeconomic policies to an NGO sector that is highly concerned about development's negative impacts on children and youth. Only recently, however, have policy-makers and practitioners begun to recognise children and youth as subjects with their own ideas, able to act in their own interests. Increasingly it is recognised that failure to listen to young people's voices has often meant failure to address many of the issues that confront children and youth.

Key ideas

- Development theories and policies have tended to neglect both the impacts of development on young people and young people's roles as actors in development.
- The main practical implications of development policy on young people relate to education and health policies and household economic status.
- The move towards NGO-led people-oriented development in the 1980s and 1990s gave greater emphasis to the needs of children and youth.
- Structural adjustment policies and economic globalisation have tended to have negative impacts on young people.
- Wealth disparities are widening globally, and affecting children disproportionately.

● Child poverty is a growing problem that needs to be understood in terms of more than income alone.

Discussion questions

- Is it better for young people if international aid is channelled through NGOs or Third World governments?

- What contributions do children and youth make to 'development'?

- Is it possible to develop a valid way of measuring child poverty that allows international comparisons to be made?

Further resources

Books

Scheper-Hughes, N. and Sargent, C. (eds) 1998 *Small wars: the cultural politics of childhood*, Berkeley: University of California Press – both the introduction and many of the chapters examine the global economic contexts of children's lives.

Stephens, S. (ed.) (1995) *Children and the politics of culture*, Princeton: Princeton University Press – addresses the global influences on the lives of young people.

Journals

Save the Children publishes a regular magazine-type journal *Children and Development* with short articles aimed at a non-specialist audience.

Websites

Oxfam http://www.Oxfam.org.uk/ has numerous online resources relating to debt and aid.

Save the Children http://www.savethechildren.org.uk/ contains reports on the impacts of a range of development policies.

UNICEF http://www.unicef.org/ and the Innocenti Research Centre http://www.unicef-icdc.org/ publish a range of relevant online documents.

Young Lives http://www.younglives.org.uk/ is an international longitudinal study of poverty in Ethiopia, India, Peru and Vietnam and a range of working papers on child poverty are available on the website.

3 Changing cultural contexts

Key themes

- Children and their families
- Children, peers and play
- Changing concepts and experiences of childhood
- Becoming an adult
- Youth culture and globalisation.

> 'Who' the child is (as with class, gender and even personality) and 'where' the child comes from (both in place and time) define important situations (or positions) from which to understand the complex and multiple realities of children's lives.
>
> (Matthews and Limb 1999: 65)

Chapters 1 and 2 dealt mainly with the international contexts in which young people grow up. Chapter 3 addresses how young people's lives are lived in, and shaped by, more local contexts. Although children and youth everywhere are faced by a degree of marginalisation on the basis of age, they are also affected by many other aspects of their lives, which may be more important to them. Individual (and collective) identities and experiences of childhood and youth are shaped by attributes which range from characteristics of the body (sex, 'race', dis/ability, age); characteristics of their families, and their position within the family (including birth order); and wider social distinctions such as class, affluence, religion, ethnicity, caste. All of these attributes gain particular meaning in different social, cultural, political and economic contexts, as well as in different physical environments. Although many contexts are adult dominated (Welti 2002), young people themselves engage in processes that are played out differently in settings that range from the very local to the global and include institutional environments (e.g. schools), workplaces, the media and peer groups.

The sections below examine how the varied and changing social and cultural contexts in which young people grow up shape both their concrete experiences and their construction of identity and aspirations.

Children and their families

Key to the local lives of young people are the relationships between older and younger members of societies, and in particular families, since, for most children, families provide their immediate contexts. A useful tool for understanding why families differ is the notion of 'intergenerational contracts', which conceptualises family relationships as a set of implicit understandings concerning the roles and responsibilities of family members. In all societies, goods and services are transferred from the 'working' generation to both the young and the elderly (McGregor *et al.* 2000). Such practices are culturally entrenched and, in transferring resources to the young, the working generation assumes that when they are old the young will reciprocate. Within this moral economy, decisions are made concerning family size and form; childrearing practices and behavioural expectations; investment in health and education; as well as expectations concerning youth transitions to work and marriage. Expected roles and responsibilities are often gendered, which contributes to unequal treatment of girls and boys (Kabeer 2000). Although parents may act on the basis of altruism or affection, '[i]n contexts characterized by high levels of uncertainty, where no institutional alternative to the family as a source of social insurance has emerged, parental decisions are likely to be powerfully motivated by concerns about their own security in old age' (Kabeer 2000: 478). Intergenerational contracts encompass not only relationships between children and their parents, but also responsibilities across extended families. Similarly, reciprocal relations may underlie transfers between children, older siblings, for instance, working to pay their younger siblings' school fees.

Intergenerational contracts differ in part because they are constructed within wider social and economic contexts. As contexts change, so do the expectations laid on different generations. In the post-war era in the West, intergenerational contracts are secured substantially at a societal level, through the state and the market (McGregor *et al.* 2000), but elsewhere most children are expected ultimately to contribute to their families' welfare. The nature and timing of children's contributions varies, however. If infant mortality declines, children are perceived to be more likely to survive and provide for their families in the future. Parents tend, therefore, to invest more in education, enhancing their children's future capacity to assist, rather than requiring immediate contributions. Any (implicit) renegotiation of intergenerational contracts in situations of socio-economic change can provoke conflicts between generations.

The meaning and value of children

The concept of childhood, and importance attached to children, is culturally constructed and therefore varies between societies, as well as between individual children within societies. Although Aries argues that childhood is a recent Western invention, most societies have some concept of childhood, albeit often shorter, particularly among the poor, than that experienced by Western children. There are, however, other differences. While Western culture views children as individuals, moving towards autonomy, many cultures construct children as fundamentally part of a family, lineage or clan. Thus negotiation of autonomy is a relatively unimportant aspect of adolescence in India, where two-thirds of young people accept their parents' involvement in finding a marriage partner, or in the Philippines, where young women migrate to the cities in order to send money home to their families (Brown and Larson 2002).

Children's importance within families is culturally varied. Among the Dinka and Nuer of southern Sudan, children are 'the reason for everything we do' (Jok 2003: 3). For a man to die without surviving children is to die a 'complete death', with no permanent position within the male ancestral lineage. In much of Africa, children represent not only lineage continuity but also the material survival of families and communities (Rwezaura 1998). They are 'wealth-in-people', expected to provide labour and support the older generation and those who are sick or needy. Children's contributions to their families may be immediate or delayed. In Egypt, for instance, middle-class parents work to control their children for future benefit, whereas lower-class parents work with their children to satisfy the family's immediate needs (Booth 2002). The image of children as a resource is, however, increasingly contested in many societies, not least by children themselves. In Africa this has often led to tensions between generations (Rwezaura 1998).

Children are valued in all societies, and in most devotion to one's children is upheld as an ideal. However, the Western cultural expectation of unconditional parental devotion, although not always translated into practice, makes it uncomfortable to recognise that emotional investment in offspring is not a universal cultural expectation. Scheper-Hughes' research in a Brazilian shanty town found many women exhibited detachment and indifference, particularly in relation to quiet, docile and inactive babies, neither christening nor even naming them until they began to walk or talk, demonstrating the risk of death was passed (Scheper-Hughes and Sargent 1998). '[D]espite the presumed universality, sanctity and immutability of parent–child bond, parent–child relationships are products of the material realities in which families are located' (Kabeer 2000: 478).

Greater value may be attached to some children than to others, and this leads to differences in their treatment (Box 3.1). Children who have a major role in the household economy or who will perpetuate the family name are in less danger of abuse than others. Birth order and gender are both important here. Those at greatest risk may be girls, the youngest in large families, children with particular behavioural or physical characteristics, second children of the same sex, those who are sick, orphaned, adopted or stepchildren or the weaker of multiple births (Boyden 1991). Neglect is most often associated with poverty, not malice – though this should not be used to stigmatise the poor.

Family types and sizes

Families vary greatly in character. They may be close knit or diffuse; hierarchical or egalitarian; extended or nuclear; female-headed or polygynous; co-resident or scattered. Children's families usually, but not invariably, include their biological parents. The roles usually performed by biological parents in the West (bearing and begetting; endowment with civil and kinship status; nurturing; training and sponsorship into adulthood (Goody 1982, cited in Tolfree 1995)) may, in other societies, be shared between several individuals. Across the world, family types are changing. Rural–urban migration and labour migration by individual family members have had considerable impacts, often leading to smaller, more nuclear households (Brown and Larson 2002).

The traditional family form in much of India was the joint family – a co-resident unit comprising a patriarch, his wife, unmarried descendants, married male descendents and their wives. As many

---

Box 3.1

Preferential treatment of girls in Jamaica

Concepts of family life among Jamaicans are dominated by Victorian ideals – both men and women tend to believe that fathers should be breadwinners and women responsible for domestic tasks, including childcare. In practice, however, men often fail to live up to the Victorian ideal through lack of employment, while female labour-force participation is among the highest in the world. Although men value fertility, they seldom show great interest in the children they father. A study in the early 1980s suggested more than 40 per cent of couples separated before the birth of their first child: many households are female-headed. Women often see men as irresponsible and unreliable, but have high expectations of their daughters. A survey found 78.7 per cent of women had wanted their first child to be a girl – only 12.8 per cent wanted a boy. Women associated the terms 'bad' and 'rude' with boys and 'loving' and 'reliable' with girls.

Both men and women think children need strict discipline, including plenty of 'flogging'. Both girls and boys do household chores. However, girls are supervised much more closely than boys – girls are expected to pass exams, find lucrative employment and contribute economically to their families. Boys play unsupervised outside much more, and are permitted to roam around the neighbourhood. This is not considered to be neglect, but does appear to correlate with nutrition indicators. Girls are consistently taller and heavier compared with boys, and the smallest boys are those who are least supervised. Theoretically, if boys and girls are treated equally, mortality rates should remain equal from 12 months until the childbearing years. In Jamaica, however, mortality among boys is higher up to the age of nine. Male children are admitted to hospital more often at all ages and experience more than twice as many accidents and injuries as girls. Furthermore, the number of boys in children's homes in Jamaica is much greater than the number of girls – there are almost twice as many homes catering exclusively to boys and boys greatly outnumber girls in mixed homes. This reflects both the higher level of abandonment of boys and people's reluctance to foster or adopt boys, who are seen as potentially difficult.

Source: Sargent and Harris (1998).

as 50 members shared both production and consumption and all major family decisions were made by the patriarch (Verma and Saraswathi 2002). In some parts of Africa, and elsewhere, polygyny remains common, although it was never universal. In some polygynous marriages the distribution of resources favours the senior wives, hence, even within a single family, children have differential access to resources (Nsamenang 2002).

Many young people grow up in families where at least one biological parent is absent. Often such families are female-headed, resulting from death, divorce, separation or abandonment, or from temporary sexual relationships or 'visiting' relationships with migrant workers that are common in some parts of southern Africa and the Caribbean (Boyden 1991). While in some contexts female-headed households are common, in others they are considered deviant. Female headship is often a strong predictor of child poverty (Fussell and Greene 2002). In Botswana, for instance, divorce is usually, but not always, associated with a declining standard of living (Maundeni 2002).

Large families are often thought to have negative consequences for children. This idea is based on assumptions that children's well-being depends on resources provided by their parents, and that those resources are fixed and shared between siblings (Desai 1995). Empirical research reveals that children who have a sibling less than five years older than themselves are, on average, shorter for their age than those who do not; those with a sibling 12 or more years older are taller. This suggests that older siblings contribute to the costs of younger siblings. However, the impact of family size differs depending on the context. Economic growth and improvements in health infrastructure benefit children in small families most (Desai 1995). Privatisation and public expenditure cuts make families more responsible for their children's welfare, disadvantaging large families and requiring a reworking of intergenerational contracts (Desai 1995). Family size is falling in many countries today, partly due to a growing prevalence of nuclear families, and partly a (related) fall in fertility rates.

Childrearing practices

The meanings attached to children, along with beliefs about the nature of childhood and desirable outcomes, shape childrearing practices. All cultures see children's development in moral terms, but there are contrasting ideas both about what behaviours should be encouraged in children and how best to facilitate their acquisition. Often childrearing is gendered: Indian girls, for instance, experience more control over their social and household activities, while boys' academic work comes under close family scrutiny (Verma and Saraswathi 2002).

Some African societies use the metaphor of a seed, nursed to maturity by a range of actors (Nsamenang 2002). The Hausa believe children are born without a definite character and that upbringing is crucial in shaping future character. This is expressed in proverbs relating to childrearing: 'character is like writing on a stone'; 'a stick should be bent when it is raw' (Schildkrout 2002/1978: 354). By about seven, however, Hausa children are said to develop understanding or sense, which they acquire through experience, and can assume responsibility for their own behaviour. Childrearing in China similarly assumes a modelling theory of learning – emphasis is placed on functioning as a desirable model for children to copy. There is a strong belief in children's natural benevolence – that children are basically good and readily malleable. Throughout early childhood parents are expected to be tolerant, but once children begin school they are deemed capable of reasoning and expected to conform to socially appropriate behaviour (Stevenson and Zusho 2002).

Development of autonomy is less important in some cultures. In Southeast Asia, children grow up accepting dependence on their families – they are not expected to develop rapidly and are not strictly disciplined (Santa Maria 2002). In Puerto Rico, parents are expected to help children whenever called upon to do so, to avoid emotional upsets, and children may remain very dependent into middle childhood (Gannotti and Handwerker 2002).

There are also cultural variations in the extent to which parents or other adults are expected to exercise authority in shaping children's characters. Setswana proverbs emphasise children's subservience to their parents such as '[a] child's parent is its god' (Maundeni 2002: 288). Parental authority is often greater in nuclear families, particularly in competitive urban environments where pressure is put on children to achieve at school (Boyden 1991). African youth generally accept the rights of their parents to deal with them strictly, accepting that it is in their best interests (Nsamenang 2002). Similarly, despite the authoritarian relationships between parents and children in Moroccan families, there is relatively little conflict (Booth 2002).

In all societies some children fail to conform to social norms. The blame for nonconformist behaviour may be cast on parents or on supernatural causes. Ultimately, however, young people themselves take on the task of meeting society's expectations. '[T]he African worldview visualizes the child as an active agent . . . adolescents are obliged to construct a gender and ethnic identity consistent with the cultural scripts and gender demands of their worldviews and economic obligations' (Nsamenang 2002: 69).

Children's contributions to their families

Children's relationships with their families are seldom one-way, nor do children wait until adulthood to contribute to their families' welfare. Very many children in Third World societies spend more time at work than at school (Myers 1999). In many societies, work is seen not as an adult domain, where children might help, but the role of every household member for the benefit of the family (Hollos 2002). In rural Africa, most young people grow up in peasant households that are productive units, with clearly demarcated roles. They are quickly integrated, for instance as carers of younger siblings. In rural Bolivia, all children above five years are expected to contribute to household survival, and their work, which includes food preparation, childcare, feeding and grazing livestock and weeding, planting and harvesting crops, can be crucial to the family (Punch 2000). Among the Pare of northern Tanzania, girls of 12 and boys of 14 are considered equal to adults in power and skill in most work, and capable of looking after themselves and the household in the absence of adults (Hollos 2002). In northern Nigeria, Moslem women who practise Purdah depend on children to earn an income. Women cook food and may invest in other commodities (detergent, kola nuts, sugar, salt, fruit) which can be carried on a tray from house to house, or traded on the streets by children (Schildkrout 2002/1978). Without children's contribution, adult gender relations would be very different.

Children's roles, too, are often gendered. An adolescent Moroccan girl expressed regret at having been born female:

> If I were a boy, I would be outdoors. . . . At the movies, taking a walk. . . . Not like a girl, as soon as it gets dark, she has to get back home and that's it. . . . It is also that the girl does a lot of housework. The boy gets up, has breakfast, and leaves. He does not care about

(a) Collecting water, Lesotho.
Source: Author.

(b) Childcare, Lesotho.
Source: Rebecca Kenneison.

(c) Herding livestock, Pakistan.
Source: Fiona Gee.

(d) Taking grain to the mill, Lesotho.
Source: Author.

Plate 3.1 *Gendered contributions to family welfare.*

anything. That is not the case for the girls. She has to do the laundry,
sweep the floor, and cook. She gets exhausted by work.

(in Davis and Davis 1989, cited in Booth 2002: 218)

In India, individuals have specific responsibilities within their
families relating to their particular positions, which change as they
grow older. Sons are obliged to care for elderly parents, brothers to
protect unmarried sisters. This creates a sense of interdependence
within households and a tendency to prioritise family interests over
personal interests (Verma and Saraswathi 2002).

Children, peers and play

Children's experiences are shaped not only by their families and roles within the home.

Peers

Most children and youth associate not only with family members but also with other young people. Peer groups differ in size, whether or not they are related, whether they are single or mixed sex, and uniform or mixed age groups. The importance of peer groups varies between societies, but in many contexts young people's peers are becoming more important in their lives, particularly as school attendance increases.

In rural and poor urban communities in many parts of Africa, groups of children ranging from toddlers to older children are often found in the care and supervision of two or more teenagers. The children grow up in this peer group, spending much more time in the company of their peers than is common in the West. Even as toddlers they spend more time, and interact more, with peers than with adults (Nsamenang 2002). In northern Nigeria, for example, Hausa children are weaned soon after they can walk and join other children, learning a great deal from them. Divorce is common and many children experience repeated changes of residence and caretaker, but their peer group may remain a relatively constant presence (Schildkrout 2002/1978).

In India, non-kin peers are much less significant in the lives of children and youth (other than among the rich and the very poor). Instead, joint families of up to 50 members provide plentiful peers under the same roof (Verma and Saraswathi 2002). Similarly, peer relationships between children in Arab societies have been mainly between same-sex cousins. With formal education, and later marriage, particularly in urban areas, peer relations are growing in importance (Booth 2002).

In contrast to families, peers are often seen as a potentially negative influence on young people. 'Worldwide there is anxiety about the negative impact of peer pressure on the young, about young people out of control, roaming the streets and indulging in crime, and about adolescent promiscuity and drug-taking' (Boyden 1991: 35). Most studies of youth peer relations focus on drugs, delinquency and illicit sex.

Play

Play and leisure activities perform important roles in the development of young people's identities and sense of belonging to a community (United Nations 2002). Everywhere children play, but the attitudes of adults towards such activities and the time available for play vary between societies. Children engage in many kinds of play: alone; with a few friends or in large groups; child-initiated or organised by adults; imaginative or rule-driven.

Children in urban India engage in varied forms of play, most involving physical activity: traditional games, pretend play such as pretending to be film stars or enacting festivals, singing games, building mud houses. Most play is opportunistic. Children use anything in their environment, including discarded objects, integrating play into their daily lives, even combining it with work,

(a) Catching birds with catapults, Lesotho.
Source: Author.

(b) Making music, Lesotho.
Source: Author.

(c) A netball match, Zimbabwe.
Source: Author.

Plate 3.2 Leisure pursuits.

whether childcare or rag picking (Oke *et al.* 1999). In rural Bolivia, children's spatial autonomy provides them with the opportunity to make time for playing. Some meet their friends on the way to school and might arrive late – or even not at all; some arrive at school early so that they can play in the square before lessons; more commonly they play on the way home or when out collecting water or herding livestock (Punch 2000). Children from more disadvantaged Indian backgrounds often play more imaginatively than those from more affluent families. Given the shortage of spaces designed for play, children use building sites, pavements, roundabouts, staircases, abandoned buildings, railway tracks, parking areas and school compounds. Many of these settings present physical dangers from, for instance, traffic, rubbish and construction machinery (Oke *et al.* 1999).

Children's play does not always have the support of parents. Play is accorded a special significance in Hindu cosmology, viewed as a metaphor for the universe, which is seen as reflecting the creation or 'play' of a cosmic power. Poets and authors have eulogised the lack of importance children attach to the outcomes of their play, the ease with which they enter and leave play and their total immersion, oblivious to their surroundings (Oke *et al.* 1999). Not all Indian parents, however, idealise play. Parents often prefer their children to be working, either for immediate benefit or, through school work, for future gain. In the Moslem Middle East, play is regarded as a waste of time (Fernea 1995), and Bolivian parents use a superstition known as the *duende* (dwarf) to deter children from playing. This frightening spirit is said to appear to children who play too much (Punch 2000). Among older and urban children, fears extend to threats of AIDS, delinquency, violence and drug abuse (United Nations 2002).

Changing concepts and experiences of childhood

Concepts and experiences of childhood are changing in many parts of the world, partly as a consequence of education and urbanisation. The 'global model' of childhood is widely represented as an ideal and is affecting children's experiences. Nonetheless, childhood remains culturally and historically specific and is never simply an exported Western institution.

The growth in formal schooling has impacted on children's day-to-day lives and has also altered the economics of childrearing, by making children a net cost to households for much longer (Box 3.2). Children's lives are also affected by urbanisation. Urban families contain individuals with differing occupations and goals, rather than a single production unit (Nsamenang 2002). Members tend to live more independently of one another and, while incomes may be shared, families seldom work together (Boyden 1991). Urbanisation is also contributing to a shift from extended families towards smaller nuclear families in many places. Eighty per cent of migrants to Indian cities settle in nuclear families (Boyden 1991).

'Global childhood' is promoted to people through the media, and by governments and NGOs. In Ecuador:

> billboard murals contrast images of children in tattered and patched clothing with dishevelled hair pushing wheelbarrows heavily loaded down with bricks and rock with those of fashionably dressed, smiling children reading books. Next to the images, bold statements proclaim 'Childhood: A Time for Studying, Playing, and Growing – Not for Working'.
>
> (Pribilsky 2001: 261)

In Singapore, it is felt that the modern Western school system has introduced the global model of the child as possessing rights: freedom from labour, the right to learn, the right to enjoy a certain standard of living and welfare and access to consumer goods. This is a much more individualistic view of the child than prevailed in the past (Yun 2000). Some Singaporean parents are enthusiastic about this rights-bearing idea of the child, but nationally concern has been expressed about defiance among the young. One manifestation of the new notion of childhood has been in young people's views of politics. A survey of 440,000 21–30-year-old Singaporeans found that 77 per cent thought there should be an opposition in parliament. Young women, in particular, are shocking society by engaging in forms of behaviour that were once the preserve of young men. Far from promoting global childhood, the government is in moral panic: '[i]f children lose respect for their elders and disregard the sanctity of the family, the whole society will be imperilled and disintegrate' (Senior minister Lee, cited in Yun 2000: 64).

In China, the government set out to change the relationship between young people and their parents in the Cultural Revolution of the 1960s. The intention was to switch young people's loyalties from their families to the state. They were urged to report their parents' behaviours, and many left home or were sent away to the countryside

Box 3.2

Changing childhoods in Ecuador

The Ecuadorian Andes have witnessed growing numbers of men undertake temporary labour migration to the USA, with considerable impacts on the lives of their children. As families become increasingly dependent on remittances from absent fathers, rather than the local farm economy, children's labour ceases to be necessary to the household. At the same time, schooling is seen as crucial preparation for children's own future migration, and carries symbolic weight in rural communities, distinguishing the children of migrants from those of impoverished non-migrants. Fathers rationalise their migration as a way of meeting their children's needs – they buy consumer goods and pay school fees. They also seek to parent in ways that more closely resemble the models represented on American television – enjoying close emotional relationships with their children, albeit separated for several years at a time. Families are increasingly child-centred, yet children are seen as more autonomous individuals, rather than part of a family production team. When migrants build new houses for their families, children are often provided with their own bedrooms for the first time, reducing the interrelatedness of families yet further. Some children appreciate the privacy this affords, but others are scared of being alone and continue to sleep in the living room or with a parent.

Many boys whose fathers are labour migrants have been diagnosed with the condition, *nervios,* characterised by depression, anger and uncooperativeness. This condition is common in adults across Latin America, and is usually attributed to breakdown of reciprocity in relationships. *Nervios* in children is interpreted by parents through classical Western theories of attachment and bonding, derived from developmental psychology, that pervade the Ecuadorian literature on childrearing. Pribilsky suggests an alternative explanation. Going to school, and receiving consumer products as gifts from migrant fathers, makes children feel indebted. Children's capacities to reciprocate are delayed, and there is no guarantee that schooling will enable them to repay their parents. 'It is perhaps the mismatched placement of childhood roles and responsibilities that is the greatest source of trauma for children in the rapidly changing communities of the Ecuadorian highlands' (p. 269). *Nervios* also serves to give social sanction to behaviour that would otherwise be condemned, bringing sympathy to children who are the objects of considerable investment by their parents.

Source: Pribilsky (2001).

(Stevenson and Zusho 2002). More recently China's one-child policy has changed experiences of childhood. A generation of children is growing up without siblings; the next generation of a society once dominated by extended families will have no aunts, uncles or cousins. Fears that the one-child policy would result in spoiled and egocentric children appeared vindicated, with preschool children labelled 'Little Emperors' and 'Little Empresses'. Although interaction with peers at school reduced children's self-centred behaviour, Chinese adolescents today prioritise their own satisfaction over their families', and ironically, many aspire to a future without children of their own (Stevenson and Zusho 2002).

Becoming an adult

Most societies conceptualise the life course from birth to adulthood as a series of stages, each with a different name and different expectations concerning the role the young person plays in their family and more widely. The number and length of stages differs considerably between societies, and sometimes between genders within a society (Table 3.1): childhood often ends sooner for girls than for boys. Very often the stages relate less to chronological age than to other attributes of the young person. In Bangladesh, an individual who attends school and has no social or economic responsibilities may be referred to as a child (*shishu*) until puberty, whereas a working six-year-old is not *shishu* (Blanchet 1996). In Indonesia, by contrast, street children refer to each other as 'child' (*anak*) well into their early twenties, and unmarried girls may be referred to as children even in their late twenties (Beazley 2002).

In societies where young people marry at puberty and move directly into adulthood there may be no term for adolescence or youth, but as societies change, the need may arise to describe a newly recognised life stage (Box 3.3). In China, people refer to whether a young person is a middle-school or a high-school student to indicate their stage in the life course (Stevenson and Zusho 2002). In India, there are no terms in widespread use, but 'youth', and increasingly 'adolescence' are used at policy levels. In Arab countries, school presents the concept of 'adolecence' to teenagers, along with the expectation of certain behaviours associated with it (Booth 2002). The term 'youth' is usually applied to young men who are more publicly visible: girls in Africa tend to marry younger and become

Table 3.1 *Different stages in the process of becoming an adult among the Sereer people of West Africa*

Boys

Name of stage	SISSIM	GAYNAK	PES	WAYABANE
	Young of the tribe	Shepherds	Young people	Youth and adults
Age	8–11	12–18	19–26	27–35

Girls

Name of stage	FU NDOG WE	NOG WE	MUXOLARE
	Young girls	Adolescents	Adults
Age	7–10	11–18	19–26

Source: Adapted from Boyden and Ennew (1997).

adult without necessarily being 'youth', although it is possible for young men to remain 'youth' even if they marry (de Waal 2002).

In some societies, a *rite de passage* marks a formal end to childhood. This may happen anywhere between the ages of 8 and 20 or older (Hardman 2001/1973). In Africa, such a rite often marks the point at which young people first take adult roles in the jural, cultural and ritual affairs of the society (Nsamenang 2002). Circumcision may,

Plate 3.3 *Boys processing home from 'circumcision school', Lesotho.*
Source: Author.

for instance, mark a boy's transition from the world of women and children to the world of men. Full adulthood, though, requires one to be married with children.

Most societies today employ age-based definitions of life stages, at least for legal and policy purposes. Many countries set a legal age of majority at 18, in conformance with the CRC. This has been widely criticised as arbitrary and at variance with prevailing expectations in many societies. Among the Tamang people in Nepal, for instance, girls marry at the age of ten, and are then considered adult, although it is only as their own children grow up that they begin to take on decision-making roles (Johnson *et al.* 1995). In Latin America, among certain (mainly rural and lower-class urban) social groups, a girl's fifteenth birthday is celebrated with a party that indicates she is ready for marriage (although in practice few marry so young) (Welti 2002). Furthermore, most countries employ a range of age-based definitions. India's Juvenile Justice Act defines a juvenile as a male under 16 or a female under 18 years (Verma and Saraswathi 2002). By contrast, the 'Youth Policy Act' in India is intended to provide services to people up to the age of 35 (Verma and Saraswathi 2002).

Youth transitions

Nowhere, then, is there a single point at which a person moves from a status of having no adult roles or rights to full adulthood. Instead, young people make a gradual transition, with changes in a number of aspects of their lives. Research in the West has focused on three broadly parallel transitions to adulthood: beginning work, leaving the parental home and beginning a new family. These were once thought of as simple linear progressions from dependence to independence, but many young people's transitions today are more complex than this, even in the West. In Third World contexts, children may begin work at a very young age, and might never be expected to leave their parents' home. Transitions make take different forms. In Africa, traditionally:

> girls would usually be married shortly after achieving sexual maturity, and consolidate their adult status when they became mothers, while boys would achieve 'adult' status by degrees, through initiation, eligibility to fight, marriage, acquisition of land, and elevation to the position of elder.

> (de Waal 2002: 14)

It is also significant that many societies do not have a concept of universal citizenship, in which every individual attains full rights. In many African communities full adult status was in the past achieved only by some elderly men and very few women (de Waal 2002).

The notion that young people become more autonomous is also problematic. In some societies youth is understood as a time when young people become progressively more integrated into the family and community, rather than becoming more autonomous. In Arab societies, children learn to prioritise the family over themselves and to feel a lifelong responsibility to their relatives. The notion of intergenerational contracts suggests young people generally move from being net consumers to net providers for their families. '[M]aturity is defined in part by how well the individual works within and maintains a web of kin and other relationships' (Booth 2002: 213). In Bolivia, parent–child relations are interdependent, with children working from a relatively young age. As they grow older, 'young people move in and out of relative autonomy and dependence' (Punch 2002b: 124), sometimes drawing on household resources, sometimes making substantial contributions.

In many societies it is marriage (rather than work or an independent home) that is seen as the most crucial landmark en route to adulthood. In many societies girls, in particular, marry young. Besides exercising control over young people's sexuality, parents may benefit from dowry or bridewealth. In some states in India, the median age at marriage is only 15 or 16 (Fussell and Greene 2002). Most young people in India are married by the age of 20, and 19 per cent of children are born to women aged 15 to 19 (Verma and Saraswathi 2002). Among the Hausa of northern Nigeria, girls ideally marry as soon as they are sexually mature: most marry between 12 and 16. At this stage a young woman enters purdah and loses the freedom she had as a child. Boys marry much later, when they are economically productive, but must move out of the sphere of women at puberty (Schildkrout 2002/1978). As formal education becomes longer, however, marriage in most societies is increasingly deferred, creating a new life stage between childhood and marriage.

Even when married, a young person may not be regarded as fully adult until they have a child. Not all young people marry, and sometimes it is childbearing that is the key moment in the transition to adulthood. A young black South African woman began relating her life story to Leonard Lerer (1998: 232) as follows: 'You start off as a child, then this boy comes along, you get this idea in your head, someone is interested in you, you have a baby and become a woman.'

Premarital sex is increasingly common in many parts of the world, particularly where age at marriage is increasing. In most places a double standard is employed. While premarital sex is disapproved of, it is tacitly recognised as more acceptable for young men than young women. Consequently, in India, adolescent boys' most frequent sexual partners are commercial sex workers (Verma and Saraswathi 2002). Attitudes differ somewhat from place to place. In the Arab world, girls are reportedly reluctant to get involved with boys, not so much because of the morality of sex itself but the violation of their fathers' trust (Booth 2002). Young women in Latin America are said to hold increasingly conservative attitudes to premarital sex (Welti 2002), whereas in South Africa, although premarital sex is usually disapproved of, and pregnant teenagers are expelled from school, premarital pregnancy does serve to demonstrate that a girl is fertile, which can make her a more desirable marriage partner (Nsamenang 2002).

Much less attention has been given to the role of work in defining adult identities in the Third World, largely because work is seldom seen as an adult preserve. The difficulties school leavers face in finding work are discussed in Chapter 6. Entering employment is a more significant aspect of transitions to adulthood where it entails migration away from the family home, although this is not always an expression of increasing autonomy. In Southeast Asia, for instance, many young women go away from home to work prior to marriage, either to the city or abroad, while young men remain at home. However, the freedom young women enjoy is for the purpose of earning money to assist their families (Santa Maria 2002). In rural southern Bolivia, children must decide at the age of 12 to 13 whether to stay at home and work or go away to secondary school, or to work in town or in neighbouring Argentina (Punch 2002b). These decisions are not made on an entirely individual basis. Structural constraints restrict young people's choices, and decisions are made through negotiations with parents concerning the potential benefits to the household, as well as the individual.

When young people move out of the family home it is not always associated with work. Some young people move when they marry. In the Philippines, more young people are leaving their family homes to establish their own households, or moving into other households, before marrying. When members of a family face difficulties, however, they tend to return to the family household for support, and there are increasing numbers of married youth residing with parents (Santa Maria 2002). In Latin America, many young people live with their parents well into their twenties, due to the shortage of both jobs and housing (Welti 2002), and, in Malaysia, young married couples often live with the wife's parents for economic reasons (Santa Maria 2002). In stark contrast, the Peruvian national census in 1981 recorded 7,000 boys and 4,000 girls aged 6 to 11 as household heads. Some may have been migrants to the cities fleeing armed conflict in the countryside; others, groups of runaways and abandoned children who set up home rather than live on the streets (Boyden 1991).

Youth transitions research in the West has focused on how young people are increasingly able (and required) to make choices in their transitions to adulthood. In many societies, transitions have been automatic in the past, young people following the route their 'culture' dictated. In some parts of the world, there is now greater divergence in the transitions undertaken by different young people, particularly where education becomes more available and marriage can be delayed. 'As longer periods of education, greater mobility, urbanization, and exposure to media become common, some Arab youth are beginning to have more say about when and how they become adults' (Booth 2002: 212). In Arab societies, however, these are mainly the wealthier and more urbanised youth, and others call for a return to traditional practices under an Islamist banner, blaming the West for a perceived breakdown of social boundaries, particularly gender boundaries (Booth 2002). Furthermore, while many young people are led (by, among other things, education) to expect that they will be able to choose their adult lifestyles, in practice the available opportunities are highly constrained (Ansell 2004).

Youth culture and globalisation

In those societies where the space between childhood and marriage is expanding and being filled by a stage that is generally labelled 'youth', young people are not only increasingly making (albeit constrained) choices concerning employment, residence and relationships, but they are also engaging in new forms of cultural expression. As a result of processes of globalisation more young people are growing up in urban environments, attending formal schools, preparing for jobs in a capitalist labour market and consuming standardised components of 'globalising' youth culture (Brown and Larson 2002). It is noteworthy that whereas literature on youth culture and subcultures in the West has focused on working-class youth, that in the Third World has explored the engagement of urbanised and relatively affluent young people with elements of Western culture. Growing commonalities may be apparent in the appearance and behaviours of young people in cities around the world, but for many experiences are in fact diverging. A 'digital divide' denies certain opportunities to poor youth, and:

> those who have little access to the trappings of this global culture, or who reject it in the face of religious or cultural mandates, may grow more estranged not only from adolescents in other nations but also from age mates in their own country.
>
> (Brown and Larson 2002: 13)

Youth cultures around the world tend to be spatially open. 'Across the world even the poorest young people strive to buy into an international cultural reference system: the right trainers, a T-shirt with a Western logo, a baseball cap with the right slogan' (Massey 1998: 123). The movement of cultural symbols is not entirely one-way, however, and the meanings attached to different commodities vary from place to place. The apparent cultural globalisation is not homogenisation. Youth cultures are local products of interactions across space: interactions that are constructed through social relations imbued with power. The social relations that underlie the wearing of a T-shirt with a US logo by a young person in Guatemala are not the same as the wearing of a Guatemalan textile wristband by a young person in the USA (Massey 1998).

The development of a variety of music known as 'Rai', which became extremely popular throughout Algeria, is an example. This is sung by young men, 'using Western instruments and mixing local popular songs and rhythms with such music forms as American disco, songs of Julio Iglesias, Egyptian instrumental interludes and Moroccan wedding tunes' (Schade-Poulsen 1995: 81). The emergence of this cultural form needs to be seen in relation to social change within Algeria since the 1970s. Rapid urbanisation has contributed to changes in family structure and greater emphasis on the couple as a unit. Youth has lengthened and women have entered the public sphere. Male youth look in two directions for inspiration in addressing their new situations: Islam and the West. Rai music, which was developed entirely by young men and spread outside any official channels, is preoccupied with women and how men should relate to them. Drawing on both Islamic and Western influences, it focuses particularly on the boundaries between gender roles that are one of the key concerns of male youth today (Schade-Poulsen 1995).

Popular Hindi films are the main youth entertainment across India. While cinema as a cultural form originated in the West, Mumbai now has the world's most prolific film industry. 'Bollywood' films are in some respects very Indian, but they are not a static representation of Indian culture, and are contributing to the production of new notions of youth 'either as romantic and dreamy or as rebellious and modern, in contrast to adults, who are portrayed as more traditional and conforming' (Verma and Saraswathi 2002: 115). Most Bollywood films are also centred on romantic love. This conflicts with traditional practices in India, and even with young people's expectations: two-thirds of young people are in favour of arranged marriages (Verma and Saraswathi 2002).

In contrast to most Indian young people, upper-class youth in Bangalore choose to disidentify with Hindi films. These young Indians create their own brand of modernity which draws on idealised images of the West and engages in an 'othering' of local Indian institutions that they regard as 'backward, hopeless and ugly' (Saldanha 2002: 345). By adopting aspects of Western fashion, music and behaviour in a near-exhibitionist way that is reminiscent of the working-class youth subcultures described in the UK of the 1960s and 1970s, they create an anti-local sense of place (Saldanha 2002).

Young people across the world use new cultural forms to challenge and subvert existing ways of doing things. This may involve selective appropriation of elements of local culture. Youth in the Solomon Islands are lured to the capital Honiara by the bright lights, and in order to escape the control exercised by rural elders over the younger generation. The term *liu* is used to refer to people without jobs who spend their time hanging around town. In town they follow customary rules only so far as these allow them some comfort in town, for instance staying with kin who cannot legitimately turn them away. As the city has only existed for half a century, there are few established norms concerning living in town. Young people

Box 3.3

Inventing teenagers in Nepal

A new magazine, *Teens*, has been marketed at upper-middle-class 10–21-year-olds in Nepal, providing them with a blueprint of what it means to be a modern teenager. The magazine's commercial rationale lies in promoting businesses that cater to an emerging market of upper-middle-class youth, such as music shops, video rental stores, beauty parlours, body-building clubs, computer and language schools, driving schools, photo studios, sports shops, yoga centres, discos and fast-food restaurants. These businesses sponsor the magazine and offer discounts to subscribers. The aim of the magazine is to constitute 'teenage' as a desired way of life that can be achieved through particular forms of consumption. Although aimed at the upper middle class, the magazine has actually attracted subscribers from across the middle classes and from a range of caste and ethnic backgrounds that would not have found a shared identity in the past. The concept of teenager appeals to a spectrum of young people seeking an identity.

The concept of teenager is widely recognised in Kathmandu (though not so much in rural Nepal), but is not associated with all young people aged 13 to 19, and is not seen in such positive terms. Parents explain that there was no such 'middle group' in the past – young people moved straight from childhood to adulthood. In Liechty's discussions with Nepali adults, '"teenagers" were always *modern consumers*: of drugs and pornography for some, and as a welcome business clientele for others' (p. 179). When used by adults, and some teenage girls, the words 'teen' or 'teenager' almost always described young unmarried males regarded as unruly or corrupt.

Through magazines like *Teens*, as well as other media, young Nepalies are presented with images of possible ways of living from beyond the nation's borders. They are encouraged to purchase their identities in the form of consumer goods, yet few can come close to affording the consumer life-styles they imagine for themselves. They are also confronted in schools and the media with an image of Nepal as a 'Least Developed Country' – a national identity that is central to the 'state modernist' ideology of progress, modernisation and development. The discourses of state modernism and consumer modernity are both highly potent: they do not coincide with lived experiences, but nonetheless impact on young people's identities. Youth in Kathmandu are prompted to refer to Nepal as 'out here' – a self-peripheralisation that reveals how central modernity has become to their conceptions of themselves, and the extent to which they feel distanced from it. Through both poverty and physical distance, young people are precluded from ever quite becoming the persons they wish to be.

Source: Liechty (1995).

are able to choose from those aspects of *kastom* that are useful to them and combine them with elements of imported Western culture (Jourdan 1995).

Although some studies represent young people as active agents in the production of hybrid cultures, other research sees youth 'as victims of modernization' (Santa Maria 2002: 171). In particular, the quest for an unattainable non-local identity may serve to alienate young people from their own societies (Box 3.3).

Religion

A factor that mediates young people's lifestyles, and particularly their attitudes to Western culture, is religion. Most young people share the religion of their families and communities. The role played by religion in the life of a child or youth varies, both with the nature of the religion and the individual's attitude towards it. The Roman Catholic church, for instance, plays a very important role in the lives of many young people in Latin America. Buddhism is highly influential in Thailand, Mongolia, Tibet, Bhutan and elsewhere in Southeast Asia. Islam is practised in the Middle East and in many Asian and African countries. While Islam arguably 'precludes total or unquestioning acceptance of "Western" lifestyles and values' (Booth 2002: 210), even in the more conservative Arab societies Arab youth do not always conform to religiously sanctioned guidelines, or see them as essential to their own faith. As with consumer culture, religion can be mobilised by young people in constructing affirmative identities, as well as a sense of community belonging, or it can inhibit the exercise of personal agency. Islamic dress, for instance, can be a source of self-expression or repression for young people. In some cases, young people either reject the religion of their parents, or react against their parents' apparent lack of religious practice. Although most Moslem youths remain moderate, some young men have adopted Islamic fundamentalism in response to Western cultural dominance and the perceived ambivalence of their parents (Booth 2002). Similarly, young people in Africa have joined Pentecostal churches in reaction against both the established churches and the gerontocracy of African 'tradition' (Box 3.4). Youth-dominated militant Islam and the Pentecostal churches are among the most powerful social movements in Africa today (de Waal 2002).

Box 3.4

Pentecostal churches in Africa

In many parts of Africa, numerous Pentecostal-type churches have emerged. There are significant differences between them, but in general they are characterised by an emphasis on personal experience of spiritual gifts and belief in 'Baptism in the Spirit' rather than set religious doctrines. These new churches, with their focus on personal inspiration, spiritual power and miracles, extreme commitment and zealous evangelism, provide young people with an opportunity to participate in a personally meaningful way. The Pentecostal churches also represent a reaction against the perceived bureaucracy, religiosity and irrelevance of mainstream African churches, and also corrupt and ineffective states and the growth of Islam. Young people are attracted to the churches' rejection of the status quo. They appear to offer both an escape from routine and a purpose to life. '[Y]oung people found and join these new churches because of the vigour of the proclamation of a relevant message and what is seen as an authentic indigenous interpretation of the Bible relevant to their religious and political aspirations' (p. 197). Although some of the churches have their origins in North America, they do not simply attract young people to a pre-constituted faith. The African Pentecostal churches are very much youth-driven and young people initiate and lead churches.

The style of worship of the new churches is exuberant and youth-orientated, with singing, dancing and popular music. Church conferences resemble rock-festivals in contexts where entertainment is short. Some of the churches are also almost explicit in their matchmaking function. Some are seen as linked to America, which increases their appeal to young people who aspire to American consumer culture. Many of the churches also preach a 'Gospel of Prosperity': it is believed that God wants every Christian to be wealthy and that true Christianity leads to prosperity. Poverty is seen as a reflection of self-inflicted sin or inadequate faith. This 'prosperity teaching' originates in North America and conflicts both with mainstream Christian teachings and traditional African cultures, which teach against accumulating greater wealth than one's neighbours.

The appeal and influence of the Pentecostal churches among young people is neither pure religious spirituality nor hedonistic consumerism. The churches are also having profound social impacts on the role of youth. They provide young people with support networks at a time when extended families are fragmenting, but they also, particularly through prosperity teaching, support those who are ambitious to break family bonds and achieve economic independence. 'It is the mass of poorly-educated youths with limited ability to gain independence who form the bulk of these new movements, which they see as offering practical and emotional alternatives to their despair' (p. 197). Such young people have little stake in the old order and embrace the focus on individualism and personal choice in challenging prevailing cultural patterns. The churches are also very popular among urban educated upwardly-mobile youth. The emphasis on personal inspiration as the true source of authority allows young people (and women) to challenge gerontocratic power. Furthermore, 'the born-again identity legitimises the actions of those youths who break from the old order and defend their independence against their elders' (p. 202).

Source: Spinks (2002).

Conclusions

Young people's lives are shaped by the immediate contexts in which they live – their families and peers – as well as by processes happening at a wider level that are translated into changing local circumstances and relationships between generations. The diversity of both ideals and realities of childhood and youth around the world raise an important question as to the extent to which it is acceptable to make judgements about the conditions in which young people live. 'Whose notions of what constitutes adversity and whose ideas about risk and developmental normality should prevail?' (Rizzini and Dawes 2001: 316). Cultural relativism is problematic, as cultural norms are by definition contested and not all social groups within a 'culture' share the same values or have the same interests – and there are differences of perspective between children and older people (White 1999). Some argue it is possible to identify basic requirements for young people to live healthy and happy lives (Box 3.5), or even that when asked, young people identify similar wishes, wherever they grow up (Malone 2001). It is also significant that young people in most parts of the world are affected by global processes, which, although they do not have the same outcomes in every context, are in many cases resulting in common patterns of change in Third World societies including trends towards smaller, more nuclear families, more time spent in school, later marriage and a growing consumer culture among young people.

Box 3.5

Pillars for good child outcomes

- A sound physical constitution;
- A nurturing family (or alternative form of family);
- A positive school or apprenticeship;
- A supportive, stable and safe community.

Source: Rizzini and Dawes (2001: 316–317).

Key ideas

- Children are usually thought of in relation to families, but families value children in different ways in different societies.
- The concept of 'intergenerational contracts' is useful in illuminating how wider social and economic contexts influence family sizes and types, and relationships between generations.
- While families generally have an important role in raising children to conform to the norms of their particular culture, many children also play important roles in their families.
- Most children relate not only to family members but also to peers.
- A range of processes including urbanisation, migration, education, global communications, religious movements and legal change are contributing to changes in both concepts and experiences of childhood worldwide.
- The ways in which young people become (and come to be thought of as) adults is highly varied, and may or may not include a period of 'youth'.
- In response to processes of globalisation, young people are constructing new 'hybrid' youth cultures, and identities that are not necessarily tied to their places of residence.

Discussion questions

- Consider the likely advantages and disadvantages for children of growing up in single-parent, nuclear and extended family households.

- In societies where young people marry in their early teens, can the concept of 'youth' have any meaning?

- Should the growing prevalence of consumer culture among youth in poor countries be seen as problematic?

Further resources

Books

Amit-Talai, V. and Wulff, H. (eds) (1995) *Youth cultures: a cross-cultural perspective*, London: Routledge – interesting research-based articles on children, youth and cultural change in diverse settings.

Brown, B.B., Larson, R.W. and Saraswathi, T.S. (eds) (2002) *The world's youth: adolescence in eight regions of the globe*, Cambridge: Cambridge University Press – a chapter devoted to each of eight regions provides a useful overview of the situations facing young people, although with an unfortunate if inevitable tendency to generalise about entire regions.

Fernea, E.W. (ed.) (1995) *Children in the Muslim Middle East*, Austin: University of Texas Press – a collection of mainly short essays about aspects of children's lives in Middle Eastern countries.

Holloway, S.L. and Valentine, G. (eds) (2000) *Children's geographies: playing, living, learning*, London: Routledge – containing several chapters relating research with children in Third World localities.

Scheper-Hughes, N. and Sargent, C. (eds) (1998) *Small wars: the cultural politics of childhood*, Berkeley: University of California Press – some chapters relate to the lives of young people in Third World countries.

Journals

Childhood frequently reports ethnographic studies with children and youth in the Third World.

Websites

UNICEF http://www.unicef.org/ publishes many online documents that detail the lives of children and young people in different places.

4 Health: ensuring survival and well-being?

Key themes

- Biomedical and alternative approaches to health
- Diseases of childhood
- Adolescent health.

The health of children in the Third World – or, at least, young children's susceptibility to disease and malnutrition – has been a dominant concern of the development community for many years. Recently, in the context of the HIV/AIDS pandemic, increasing attention has been paid to the health of teenagers. The health of a country's children is considered both a goal and an indicator of development. Health is important in relation to young people's immediate well-being: illness diminishes quality of life and poses a risk to life itself. Health also impacts on other aspects of life: young people whose health is poor are less likely to attend school or to enjoy leisure pursuits. Furthermore, ill health in childhood and youth can have impacts that last throughout life.

This chapter begins by considering different approaches to health. It then addresses the two main areas of young people's health that have received greatest attention: diseases of early childhood and health risks among those described by health professionals as 'adolescents'. In both cases attention is given to the immediate contexts of health and the wider international agendas and national policies that frame experiences of health and illness. Inevitably, many aspects of young people's health are missing from the chapter. These include young people's experiences of disability and the indirect effects on young people of the health of their relatives, especially parents, both of which receive attention in Chapter 7.

Biomedical and alternative approaches to health

Health is a complicated concept in all cultures and is not easy to define. The WHO defines health as '[a] complete sense of physical, mental and social well-being and not merely the absence of disease' (WHO 1947). This draws attention to the distinction between health as subjectively experienced by individuals and disease as an 'objective fact', defined and diagnosed by health professionals. There are cultural differences in the way ill health is defined, diagnosed and dealt with in different societies. Prominent globally, however, is the biomedical, or allopathic, approach. This has been very effective in reducing morbidity and mortality globally, particularly among the young, but by focusing on diseases rather than the people affected by them, it tends to neglect the wider social, cultural, political and economic contexts in which they occur.

Biomedicine is a highly reductionist tradition. Disease is believed to be located within the physically bounded individual, either in the mind or the body, which are seen as discrete entities. Distinct physical symptoms are viewed as indications of specific physical causes and treatment, usually with a chemical preparation, is often focused on the cause rather than the whole person.

While biomedicine seeks cures for individual symptoms, social research undertaken in relation to biomedically defined disease also focuses on individual causality. Large-scale social surveys seek causal connections between environmental and behavioural factors and the incidence of disease. 'Risk factors' are identified that appear to explain why some children and young people are more vulnerable to ill health than others.

Owing to the dominance of biomedical concepts of health, minimal attention has been given by researchers or policy-makers to subjective experiences of ill health, particularly among children. Even though physiological experiences of illness may be common cross-culturally, the ways people express and give meaning to distress depend on their cultural context (Boyden and Gibbs 1997).

Few studies have looked beyond immediate mechanisms of causality to the wider political, economic and cultural contexts that shape people's experiences of health. The dominance of a 'risk factors' approach has been criticised by those who see ill health to be related more to structural conditions of poverty and marginalisation than individual decision-making. As Lock and Scheper-Hughes (1990) comment: '[t]he illness dimension of human distress is being medicalized and individualized rather than politicized and collectivized' (cited in Boyden and Gibbs 1997: 13).

Western biomedicine is but one of many medical traditions. Many others, such as the 2,000-year-old Chinese medical system, depend on very different principles from Western biomedicine. Many medical systems are more holistic, interested in the whole person, and their interconnections with society and with the spiritual realm. This has implications for treatments: while physical remedies may be prescribed for physical illnesses, the body may also be treated for spiritual or social purposes. Equally, spiritual solutions may be sought to physical problems.

In many parts of the world, health-seeking behaviour involves negotiations between different therapeutic models and systems (Boyden 2000). Allopathic medicine may be favoured for some purposes: it is widely perceived to have brought some important advances, including declining child mortality rates. In other instances alternative therapies may be considered more appropriate. In Zambia, for instance, traditional treatments may be sought in serious cases of malaria where children suffer convulsions (Baume *et al.* 2000). While there are many documented cases where traditional remedies have proven more effective than Western medicine (at least in terms of their own intentions), the use of alternative therapies can be problematic (Box 4.1) and has been blamed for deaths from diseases for which allopathic medicine offers reliable cures. In Bangladesh, for example, research suggests that whereas child mortality rates from acute lower respiratory infection declines with increased access to (even untrained) allopathic practitioners, increased access to indigenous practitioners is associated with increased mortality (Ali *et al.* 2001).

While alternative therapies may be problematic, the biomedical focus of policy interventions has also left gaps in our knowledge and understanding of young people's health. Until the 1990s, the focus was very much on children under five: the group perceived as most vulnerable to physical disease and most susceptible to interventions.

Box 4.1

Problematic traditional treatments

Tea-tea *and* Ebino

These are physical procedures carried out on children's bodies in northern Uganda, intended to cure or prevent physical problems, but understood in relation to different causal processes from those acknowledged by allopathic medicine. *Tea-tea* consists of making cuts on the chest wall of a child who is experiencing difficulty breathing, in order to remove 'millet grains' (adipose tissue, in allopathic terms) that are deemed to be causing the illness. *Ebino*, or 'false teeth', consists of the extraction of canine tooth buds from infants because they are believed to cause childhood diseases including fever and diarrhoea. This has been a widespread practice in Uganda since the 1970s and has spread to neighbouring countries. Research in the paediatric ward of a northern Ugandan hospital found that complications arising from these two procedures, in particular septicaemia related to *ebino*, were among the leading causes of both admission to hospital and hospital death. *Tea-tea* is usually performed with a new razor blade and its dangers are less severe (Accorsi *et al.* 2003).

'Spirit children'

In some parts of Africa, certain children become defined as 'spirit children'. In some communities in Nigeria, for instance, twins are killed at birth because they are viewed as a bad omen for a family; in parts of Sudan, twins are believed to be birds and are left in baskets in treetops; in many African communities, children born with physical deformities, perceived as evil spirits, are left to die. In communities in remote northeastern Ghana, almost 15 per cent of the deaths of children under three months of age have been attributed to a belief in spirit children which resulted in infanticide (Allotey and Reidpath 2001).

Female genital cutting

In many East African countries, as well as elsewhere in Africa and the Middle East, girls are subjected to varying degrees of genital cutting. This is seen as a treatment with a social function (reducing the likelihood that girls and women will engage in extra-marital sex), and also to have religious sanction (it is associated with Islam, although it has been practised in parts of Africa since pre-Islamic times). Worldwide, 130 million women and girls have experienced genital cutting and two million girls undergo the procedure every year (WHO 2002a). While it is usually practised on girls from the age of four, it may be carried out on adolescent girls, sometimes immediately prior to marriage (WHO 2002a). In some parts of Africa, it is now in decline. In Egypt, for instance, although most girls are still 'circumcised', they are 10 percentage points less likely to be than were their mothers. Furthermore, for the majority of Egyptian girls today the procedure is performed by a physician or a nurse within a Western medical setting (El-Gibaly *et al.* 2002).

Young children were the ideal subjects of biomedicine: seen as docile bodies to be acted upon (Aitken 2001). Once the rapid increase in HIV infection among 15–24-year-olds became apparent, the attention given to the health of older children and youth (defined in biological terms as 'adolescents') dramatically increased. WHO established a Department of Child and Adolescent Health and Development, and in 1997, with UNICEF and the UN Population Fund (UNFPA), set out a Common Agenda to promote the health and development of young people aged 10 to 19 (WHO/UNFPA/ UNICEF 1997). As with infant health, the focus has been on identifying and addressing individual risk factors.

Children in middle childhood have been largely neglected in both research and policy. Among children aged 5 to 14, helminth infections such as schistosomiasis are a particular problem. These are caused by intestinal parasites that damage health and nutritional status and exacerbate the problems caused by other illnesses (WHO 2002c). Research in Somalia also found that 5–12-year-olds were more vulnerable to malnutrition than younger children, yet the health problems of this age group remain rather hidden (Boyden 1991). There is also little information available concerning children's mental health in Third World countries (White *et al.* 2002). This in part reflects the low status of mental health concerns among biomedical researchers, but also the fact that while bodies differ relatively little geographically, psychology differs much more. The validity of Western psychiatric models across societies is increasingly questioned (Boyden and Gibbs 1997).

Diseases of childhood

Young children are by far the most vulnerable age group in relation to illness (morbidity) and death (mortality). Many babies die within a month of birth (neonatal mortality). Worldwide more individuals die in the first year of life (infant mortality) than any other. Although risks diminish after the first year, children under five suffer high death rates (child mortality). Illnesses of childhood also have significant immediate and longer-term consequences for those who survive.

Six diseases (malnutrition, acute respiratory infection (ARI), diarrhoea, malaria, measles and AIDS (Box 4.2)), between them, and often in combination with one another, kill over ten million children

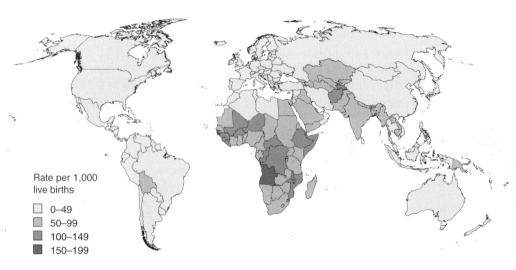

Rate per 1,000
live births

☐ 0–49
■ 50–99
■ 100–149
■ 150–199

Map 4.1 *Global infant mortality rates.*
Source: Data from CountryWatch (2003).

a year (Figure 4.1). These diseases have been the main focus of both research into children's health and international efforts to reduce infant and child mortality. Most are seen as susceptible to technical solutions – immunisation, simple drug treatments, breastfeeding – but the fact that 99 per cent of these deaths in 2000 occurred in low-income countries (see Map 4.1) suggests that poverty plays the most significant role (WHO 2002c).

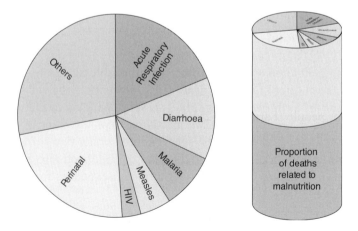

Figure 4.1 *Major causes of death among children under five worldwide, 2000.*
Source: Adapted from WHO (2001b).

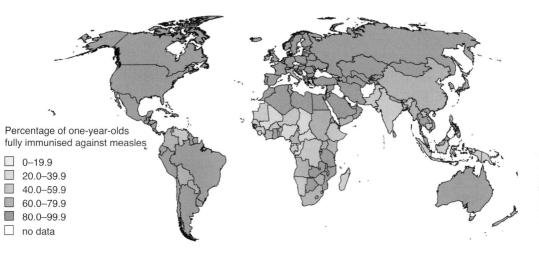

Percentage of one-year-olds
fully immunised against measles

☐ 0–19.9
☐ 20.0–39.9
☐ 40.0–59.9
■ 60.0–79.9
■ 80.0–99.9
☐ no data

Map 4.2 *Immunisation of one-year-olds against measles, worldwide.*
Source: Based on data from UNDP (2003).

Household-level practices and risk factors

Explaining the persistence of immensely high death rates from childhood diseases, despite the apparent simplicity of treatment and prevention, requires more than technical understanding. Much research has started from the position that in all societies some children succumb to disease while others do not. Studies have examined differences between and within households, and sought explanations relating to the environment and, in particular, the behaviour of household members. 'Risk factors' have been identified and informed policy responses. Such studies, however, have serious limitations: they are generally based on a narrow conceptualisation of ill health; they infer causality from data that merely indicate some degree of association between characteristics; and by inference they cast the blame for diseases of poverty on the behaviour of individual poor people (Farmer 1992).

Differential treatment of children

Children within a household do not necessarily share equal chances of good health or even survival. Globally, mortality rates among boys exceed those among girls at every age. In most countries, boys also suffer higher rates of malnutrition than girls (WHO 2000). Two exceptions are India, where girls' nutritional status is considerably

Box 4.2

The major childhood killer diseases

Malnutrition

Malnutrition is a factor in more than half the deaths of children under five (Wagstaff and Watanabe 2000). Beyond those who die, 167 million malnourished children are caught in a cycle of poverty and ill health (Smith and Haddad 2000). Malnutrition renders children vulnerable to infection, and infectious diseases undermine children's nutritional status (WHO 1998a). Children who are malnourished suffer up to 160 days of illness each year (WHO 2003).

Causes   Malnutrition is not a single disease, but a number of diseases directly attributable to deficiency (or excess) of particular nutrients. These may be micronutrients (vitamins and minerals, especially Vitamin A, iron or iodine), protein or energy. Protein–energy malnutrition is considered the most lethal form of malnutrition and is reflected in children being stunted (short for their age), wasted (low weight for height) or underweight (low weight for age).

Treatment   A diet that includes the necessary nutrients.

Prevention   Breastfed infants are less likely to be malnourished, as they obtain necessary nutrients from their mothers. Growth monitoring is seen as an effective way of alerting the parents of children who are vulnerable to malnutrition. Feeding programmes.

Acute respiratory infections (ARI)

ARI, most commonly in the form of pneumonia, kills over two million children under five every year. Up to 40 per cent of children taken to health clinics in Third World countries have ARI.

Causes   Exposure to indoor air pollution (through the burning of coal, wood, dung or fibre residues for cooking and heating, and dust in mud-built homes, combined with poor ventilation) increases the risk of pneumonia (WHO/World Bank 2001). In many cities outdoor air pollution also poses a threat.

Treatment   Low-cost antibiotics are usually effective, but as ARI can kill very quickly, treatment is urgent (WHO 1998a).

Prevention   Improved ventilation; keeping children away from smoke; improved nutrition; measles immunisation; Hib vaccine (WHO 2001a).

Diarrhoea

Two million children a year die from diarrhoeal diseases.

Causes Diarrhoea can be caused by a range of infections, most of which are transmitted through contaminated food or water. Diarrhoea is often accompanied by malnutrition (as both cause and consequence).

Treatment The main danger is dehydration, so fluid is needed. As children with diarrhoea quickly lose nutrients and energy, food intake should be normal or increased. oral rehydration therapy (ORT), a mix of clean water, sugar and salt, or oral rehydration salts (ORS), low-cost sachets of essential nutrients to be added to water, are easily absorbed and are believed to have saved many children's lives since the 1980s.

Prevention Breastfeeding of infants; improved complementary feeding; clean drinking water; sanitation; improved personal and domestic hygiene; measles immunisation (WHO 1999).

Malaria

About 700,000 children die annually from malaria, mostly in sub-Saharan Africa (WHO 1998a). Young children are more vulnerable than adults as they have not had time to acquire immunity from repeated exposure.

Causes Malaria is caused by a parasite transmitted to humans by the Anopheles mosquito which breeds on standing water in warm conditions.

Treatment Medicines are effective if administered quickly.

Prevention In recent years, mosquitoes have developed resistance to the insecticides used to eradicate them, and the parasite itself has become resistant to many anti-malarial prophylactics. Apart from drugs, the risk of malaria can be reduced by eradicating standing water and using bed nets.

Measles

Despite immunisation campaigns, measles continues to kill 800,000 under-fives annually, and a further 40 million are infected (WHO 1998a). Those who survive measles may suffer longer-term problems, including increased vulnerability to other illnesses.

Causes Measles is a viral infection carried in the air.

Treatment There is no specific treatment for measles, although symptoms such as fever can be relieved.

Prevention Vaccines against measles are cheap and effective, but many children are not immunised (see Map 4.2).

AIDS

Although AIDS is not among the top five childhood killer diseases globally, in Africa numbers are rising rapidly. An estimated 500,000 African children died from the disease in 2003, and 700,000 infants became infected (UNAIDS 2003).

Causes Where a mother is HIV-positive, the risk of mother-to-child transmission is 15–30 per cent before or during birth and a further 15–30 per cent due to breastfeeding. Some children are infected in medical procedures, for instance through the use of unsterilised needles in vaccination.

Treatment Few HIV-positive infants survive: 80 per cent die before they are five (Bakilana and de Waal 2002). Treatment is possible but, unlike treatments for the five diseases above, currently very expensive.

Prevention Mother-to-child transmission can be dramatically reduced by giving a dose of Nevirapine to mothers in labour and another to their newborn child. Breastmilk substitutes reduce mother-to-child transmission and are recommended except where access to clean water is impossible. Use of babymilk formula is, however, problematic for some mothers as it is expensive and may advertise their HIV status (Bakilana and de Waal 2002).

worse than boys' (WHO 2000), and China, where the survival chances of girls have deteriorated relative to boys since the implementation of the one-child policy. Besides gender differences, short birth intervals put some children at risk, and birth order has an impact on survival chances (Lachaud 2004). Research in India found children born less than 18 months after a preceding child were at greater risk of dying before they were two, a result of rivalry for resources (Whitworth and Stephenson 2002). Passive infants are most at risk of undernutrition, and in some cases discriminated against. The Basotho in southern Africa have a saying that 'a child that does not cry dies on its mother's back'.

Pregnancy and childbirth

The characteristics of mothers and circumstances surrounding birth have consequences for children's well-being, both in the neonatal period and subsequently. Declines in neonatal mortality have been much smaller than for older children – as many as half of those who die in infancy succumb within the first month (WHO 2002c). Maternal age can affect survival chances, with mortality rates higher where mothers are young (Folasade 2000). The mother's health, too, is significant: children born to unhealthy mothers are more commonly underweight and vulnerable to illnesses (WHO 2002c). Opinion on whether babies are likely to be healthier if born at home or at a healthcare facility is divided. There is a consensus, however, that the presence of a trained birth attendant is advantageous (Lachaud 2004). Nonetheless, 50 million women each year deliver without a skilled birth attendant (WHO 2002c).

Breastfeeding and weaning practices

'Children who are not breastfed are almost six times more likely to die by the age of one month than children who receive at least some breastmilk' (WHO 2002c: 14–15). The WHO recommends exclusive breastfeeding for six months, mainly because it reduces the risk of diarrhoeal infections in babies (Figure 4.2), and continued breastfeeding until at least two years of age in order to meet infants' nutrient needs (Dewey 2003). It is believed that a modest increase in breastfeeding could prevent up to 10 per cent of deaths of children under five (WHO 1998a). Yet there are many constraints on breastfeeding. Despite an International Code of Marketing of Breastmilk Substitutes, infant formula is promoted as a clean, healthy and modern product: women are 'turned into consumers of a commercial product that they do not need, and cannot afford, and that contributes to the death of their infants' (Scheper-Hughes and Sargent 1998: 5). The recruitment of mothers into wage labour also

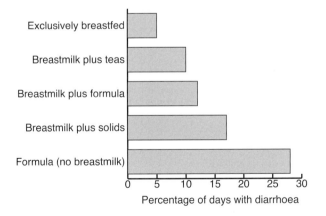

Figure 4.2 *Prevalence of diarrhoea by feeding mode for infants aged 3–5 months, Peru.*

Source: Adapted from WHO (1999).

inhibits breastfeeding. Those unable to purchase enough powdered milk may substitute sugar water and rice or sugar water and fine manioc, with even less nutritional value (Scheper-Hughes and Sargent 1998). Weaning practices, too, have been given attention, as in some places the food that babies receive is deemed inadequate for healthy development.

Local environmental health

Numerous diseases are spread through contaminants, to which children may be exposed due to factors over which their families have no control. Globally 1.7 billion people lack safe water and 3.3 billion lack access to adequate sanitation (Mehrotra *et al.* 2000).

Plate 4.1 *Environmental health risks, Kenya.*

Source: Author.

Some households are located in areas affected by domestic or industrial contamination or exposed to animal waste. It is often the poor who are compelled to live in the least healthy environments. In northern Pakistan, for instance, children suffer diarrhoea less frequently in households that have sufficient income to afford less crowded living space, separate rooms for food preparation, houses with cement floors and screened windows, separate accommodation for animals and in some cases flush toilets (Halvorson 2003).

Domestic hygiene practices

Children also encounter contaminated food and water within the home or, if children accompany parents to work, in the workplace. Hygiene can be improved through, for example, appropriate disposal of domestic waste, eating from a table rather than the floor and care in the collection, transport and storage of water (Folasade 2000). Hygiene practices may, however, be constrained by lack of knowledge or by other factors. Boiling of water, advocated for prevention of diarrhoea, requires time, energy and fuel: scarce commodities in some households (McLennan 2000). Research in northern Pakistan attributed differences in children's vulnerability to diarrhoea to mothers' access to tangible resources (for example income to buy soap, covered water and food storage containers, basins for washing babies' nappies), and intangible resources (human capital, notably education, and social capital, particularly relationships with other women within and beyond the household) (Halvorson 2003).

Management of illness in the home

Most illnesses are treated within the home, at least in the first instance. Such treatment varies from traditional remedies to the administration of Western pharmaceuticals. It is important that treatment is started promptly, but also that medicines are used correctly. A study in Zambia found many parents administering chloroquine to children with malaria were underdosing, overdosing or not completing the full dose (Baume et al. 2000): practices that contribute to the development of drug-resistant strains of the parasite. Equally, research in China suggests over a third of children had received antibiotics administered inappropriately by their parents over the preceding year (Bi et al. 2000). Use of antidiarrhoea drugs or antibiotics for treating watery diarrhoea can prove dangerous, but is encouraged by medicine sellers (Iyun and Oke 2000). While most families recognise that diarrhoea should be treated using fluid to prevent dehydration, this is contingent on access to potable water. Communities in remote areas of southwest Nigeria hesitate to treat children with oral rehydration therapy (ORT) because their water is saline and contaminated by offshore oil drilling (Iyun and Oke 2000).

Care seeking

The likelihood of parents seeking outside care when their children are ill varies considerably from place to place. In Sri Lanka, despite high rates of malnutrition, child mortality rates are low. De Silva *et al.* (2001) attribute this to the fact that outside care is sought in 65 per cent of illness episodes where symptoms of the five major childhood illnesses are present. Research in northeast Brazil, in contrast, suggests that delay seeking medical care contributes to over 70 per cent of children's deaths from diarrhoea or ARI (Terra de Souza *et al.* 2000). Reasons for differences in care-seeking behaviour are varied. The poor are often disadvantaged in availability, accessibility, affordability and quality of health care (Box 4.3). In Kerala, where allopathic care is less available in rural areas, rural families are more likely to seek care in the alternative system (Pillai *et al.* 2003). In northeast Brazil, mothers sometimes delay seeking medical care because the transportation required is unavailable or too expensive (Terra de Souza *et al.* 2000). Last, parents might not attribute their children's suffering to the pathogenic

---

Box 4.3

Characteristics of health services that affect child health

- Availability of health services (e.g. presence of a health worker trained in childhood conditions);
- Accessibility (distance, opening hours, availability of trained staff and drugs);
- Affordability;
- Perceived quality.

Source: WHO/World Bank (2001).

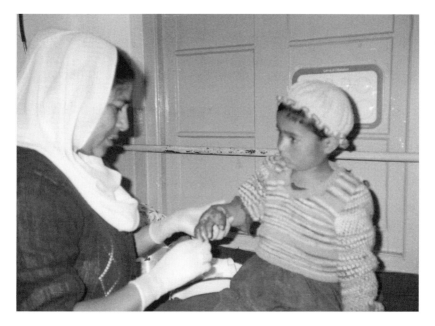

Plate 4.2 *Receiving treatment in the allopathic healthcare system, Pakistan.*
Source: Fiona Gee.

causes that are regarded as amenable to Western medicine, or simply distrust it. Some Brazilian parents fail to seek medical care for children they believe are doomed, or have experienced the 'evil eye' (Terra de Souza *et al.* 2000).

Household resources: wealth, education and decision-making

The sections above have suggested reasons why some households adopt practices that are less favourable to their children's health than others. Often the options available to caregivers are constrained by time, money, knowledge and access to facilities, as well as by their own health. One of the clearest influences on household level health care is poverty (Figure 4.3): 'in some countries children in the poorest third of the population are six times more likely to die before the age of five years than those among the richest ten per cent' (WHO 2002c: 1).

Another highly significant factor in children's health and survival is parental education. In Ethiopia, for instance, mothers' and fathers' education levels are found to have strong (and independent) impacts on child survival (Kiros and Hogan 2001). Commonly, however, it is

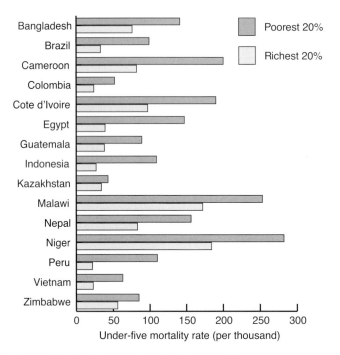

Figure 4.3 *Wealth gaps in under-five mortality.*

Source: Based on World Bank (2003).

maternal education that has the greatest impact. In Pakistan, a very strong negative association exists between lack of maternal schooling and infant survival (Agha 2000), and, in Mozambique, healthy growth of children is clearly associated with maternal education (Pfeiffer *et al.* 2001).

There are several reasons why maternal education has this impact (Figure 4.4). First, education provides women with knowledge: health knowledge in particular appears to affect child morbidity and mortality rates in some contexts (Kovsted *et al.* 1999). Second, educated mothers are more likely to engage in paid work. Increased household income generally improves children's welfare, and, compared to men, women spend more of their earnings on their children. Research in Ethiopia, for instance, found married women were willing to pay more for malaria prevention for their families than were their husbands (Lampietti *et al.* 1999). Mothers' income is likely to improve children's nutritional status and medical care. Engaging in paid work may, however, reduce the time mothers spend with their children, and thus diminish the quantity and possibly the quality of childcare, including breastfeeding. Some mothers take their

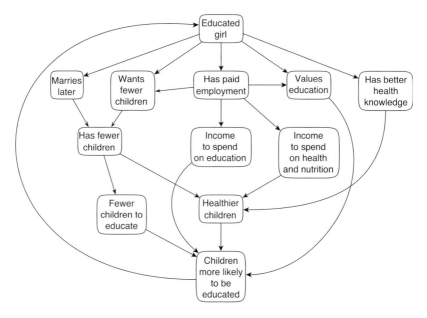

Figure 4.4 *Generational impacts of educating girls.*
Source: Adapted from Mehrotra *et al.* (2000).

children to work with them, which has both advantages and
disadvantages. Research in Nigeria suggests that children of working
mothers are less often stunted than those whose mothers do not
work; wasting was found to be less where mothers took their
children to work than where they left them at home (Ukwuani and
Suchindran 2003). Even where educated women are not employed,
however, they may have more power to influence how household
income is spent.

Wider contexts: the international and national arenas

There are geographical differences in children's health within
countries. Rates of stunting, for instance, are about 50 per cent
higher among rural than urban children (Wilkinson 2000b), although
this overall picture disguises the stark differences between urban
neighbourhoods (Boyden 1991). The situation of those in poor urban
neighbourhoods can be worse than in rural areas. The most marked
health differences, however, are those between countries. Of the
children under the age of five who died in 2000, 99 per cent lived in
low-income countries (WHO 2002c). These broad differences in
children's health can readily be explained by reference to global

Table 4.1 *Global health statistics*

	Under-five mortality rate			Infant mortality rate		
	1960	*2001*	*decline (%)*	*1960*	*2001*	*decline (%)*
Sub-Saharan Africa	253	173	32	152	107	30
Middle East and North Africa	250	61	76	157	47	70
South Asia	244	98	60	148	70	53
East Asia and Pacific	212	43	80	140	33	76
Latin America and Caribbean	153	34	77	102	28	73
CEE/CIS and Baltic States	103	37	64	78	30	62
Industrialised countries	37	7	81	31	5	84
Developing countries	223	89	60	141	62	56
Least Developed Countries	278	157	44	170	100	41
World	197	82	58	126	57	55

Source: Based on UNICEF (2003d).

inequalities in economic resources (Table 4.1). Mortality rates are much higher in poor countries, and among the poor in all countries. Under-five mortality remains 25 times higher in sub-Saharan Africa than in industrialised countries (Mehrotra *et al.* 2000).

Infant and child mortality have fallen in recent decades. Child mortality globally decreased from 93 per 1,000 live births in the early 1990s to 83 per 1,000 live births in 2000, but failed to meet the pledge of the 1990 World Summit for Children to halve child mortality by 2000 (UNICEF 2001b). Furthermore, it is the high- and middle-income countries that have seen the greatest improvements in health, while in many of the poorest countries children's health has improved relatively little. The proportionately greatest declines in infant and child mortality have been in the industrialised countries, the smallest in sub-Saharan Africa. Partial explanations of these trends lie in two contradictory sets of processes promoted by global institutions: UNICEF and WHO on the one hand and the IFIs on the other.

UNICEF and the WHO

UNICEF launched its 'child survival and development revolution' in the 1980s, promoting growth monitoring, ORT, breastfeeding and immunisation (known by the acronym GOBI) as ways to reduce infant and child mortality rates. By the end of the decade, it estimated that 12 million children had been saved. The WHO has also played a key role in agenda setting and funding research and interventions aimed at improving children's health. Continued emphasis on children's health is supported by Article 24 of the CRC, which elaborates 'the right of the child to the enjoyment of the highest attainable standard of health', and includes commitments to reduce infant and child mortality and to combat disease and malnutrition (United Nations 1989). Since the 1980s, new initiatives have been pursued, including in relation to immunisation (Box 4.4). A new approach to childhood diseases known as integrated management of childhood illness (IMCI) was developed by WHO in collaboration with UNICEF, in recognition that a simplistic division of childhood illnesses into discrete categories (e.g. ARI or diarrhoeal) does not always correspond to the combinations of symptoms children experience (Heuveline and Goldman 2000). Children taken for treatment are often suffering from more than one condition: research in Malawi found that 96 per cent of the children who met the clinical definition of pneumonia also met that for malaria (WHO 1999). To improve care for sick children and to avoid the duplication of effort that arises when diseases are addressed by separate control programmes (WHO 1998a), IMCI addresses childhood illnesses at three levels: the health system; health workers' skills; and family and community practices (WHO 1998a). Standard guidelines are provided for health workers to help them assess sick children, decide on a course of treatment and offer advice to parents (WHO 1999).

The international financial institutions and structural adjustment

The failure to make greater strides in improving children's health over the past two decades is commonly attributed to structural adjustment policies imposed on Third World governments by the IFIs. These were inspired by neo-liberal economic theory and included removal of subsidies and price controls causing increases in food prices; closure or privatisation of many sources of employment, leading to unemployment and falling incomes; and introduction of user charges for health care and education. The consequences for children's health seem predictable: reduced expenditure by poor households on health measures; deteriorating nutrition, as households reduce food intake or switch to cheaper calorie sources containing

Box 4.4

Immunisation

In 1974, UNICEF launched its Expanded Programme on Immunisation (EPI) to promote vaccination for six childhood diseases: diphtheria, measles, pertussis (whooping cough), poliomyelitis, tetanus and tuberculosis. While in 1974 the global immunisation rate for children for these diseases was five per cent, it increased to 80 per cent of children by 1990. By 1999, however, it had fallen to 74 per cent, even though the cost of protection is only US$17 per child. In some countries as few as 30 per cent of children receive the six vaccines; in Africa as a whole, fewer than 60 per cent are inoculated against measles; and globally nearly two million children still die each year of vaccine-preventable illnesses (WHO 2001a).

In 2000, the Global Alliance for Vaccines and Immunisation (GAVI) was launched as an international coalition including governments, international organisations, NGOs and the private sector. The intention is to reverse the decline in vaccine coverage and develop new and better vaccines (WHO 2001a). Many vaccines currently exist, besides the EPI six (Figure 4.5), that could reduce mortality and morbidity were they available in countries where they are needed. There is also a perceived need for immunisation against other diseases, including diarrhoeal diseases, malaria, respiratory infections and HIV, as well as safer and simpler vaccines (fewer doses, needle-free) (WHO 2001a). Immunisation not only protects the individual child, but helps protect other children who have not been immunised, by preventing the spread of disease, and is most effective when coverage is near universal. Complete eradication of a disease proved possible in the

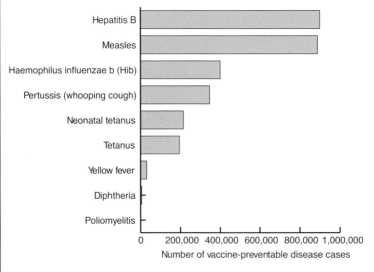

Figure 4.5 *Estimated potential impact of immunisation on vaccine-preventable diseases.*

Source: Based on WHO (2001a).

case of smallpox. If polio is eradicated by 2005, as the WHO intends, US$1.5 billion a year will be saved on immunisation costs (WHO 2001a).

Immunisation is not without critics, however. The reuse of non-sterilised syringes, in part due to insufficient supply, is estimated to transmit hepatitis B to hundreds of children a year in northwestern China (Murakami *et al.* 2003). Although vaccines are effective in preventing disease-specific morbidity and mortality, they possibly do not cut all-cause mortality.

less protein or micronutrients; and declining utilisation of mother and child health services due to user charges. It is not easy, however, to measure objectively the health impacts of structural adjustment for a number of reasons:

- adjustment is not a single, consistent set of policies;
- it is impossible to know for certain what would have happened in the absence of these policies;
- some effects of adjustment are difficult to quantify, such as the increase in demands on a mother's time;
- there are time lags involved – health status may not deteriorate immediately, and there may be inter-generational effects;
- data on health outcomes is often inaccurate or out-of-date.

(Costello *et al.* 1994)

What data there are suggest that in several countries child mortality rates in the early 1990s were substantially higher than projected from previous data; that maternal mortality rates rose in adjusting countries, especially in sub-Saharan Africa; and that the nutritional status of children and mothers declined over the 1980s in many adjusting countries, particularly in sub-Saharan Africa (Costello *et al.* 1994). Household-level research in Cote d'Ivoire suggests that the reduced availability and quality of healthcare services and increased food prices brought by adjustment had detrimental effects on children's health, manifested in both stunting and wasting (Thomas *et al.* 1996).

National priorities, policies and practice

Some countries are severely limited in what they are able to spend on health care. Forty-one highly indebted countries spend less than $10 a person per year on health care, 20 per cent below the minimum recommended by the World Bank (Simms *et al.* 2001). Not all poor countries, however, have equally poor records in relation to children's health. Children living at the dollar-a-day poverty line in different countries have very different survival chances: in Niger, 352 in 1,000 such children die before their fifth birthday; in Kazakhstan, the risk is 39 per 1,000. Similarly, in India, 58 per cent of poverty-line children are underweight, while the corresponding proportion in Kazakhstan is 12 per cent. Wagstaff (2003) attributes such differences to variations in per capita health spending. In Latin America, primary and secondary healthcare development explains sharp declines in infant mortality in certain countries and regions (Bahr and Wehrhahn 1993). Early investment in health care enabled Cuba and Vietnam to reduce child mortality: under-five mortality rates in Cuba are similar to those in Portugal although per capita income is ten times lower and, whereas Vietnam and Haiti have similar levels of income, the under-five mortality rate in Haiti is three times that in Vietnam (Mehrotra *et al.* 2000).

Although research suggests public investment in health improves outcomes for children (Box 4.5), the neo-liberal IFIs have encouraged governments to seek ways of financing health care without increasing public spending. In the 1980s, the favoured approach was user charges, intended to raise revenue and deter unnecessary use of services. User fees, however, contribute minimally to healthcare costs – as little as five per cent in most African countries – but have led to declining use of maternal and other health services in the poorest communities and increased infant death rates (Simms *et al.* 2001).

Box 4.5

Characteristics of countries that have achieved good results in social development, relative to their income levels

- The state has played a leading role;
- High priority has been given to basic healthcare and education;
- Neo-liberal policies have been avoided;
- Synergies have been sought in the social sector, e.g. between health and education;
- Policy has moved away from welfarist approaches to approaches that involve people as active agents of change.

Source: Mehrotra *et al.* (2000).

More recently, governments have been encouraged to reduce public healthcare costs by increasing private insurance cover. Where such policies have been pursued, research suggests care seeking has increased little (Thind and Andersen 2003), and any associated fall in child mortality has been minimal (Dow and Schmeer 2003). Private sector healthcare provision commonly fails to meet the needs of the poor: private providers are reluctant to set up in areas of greatest need (Thind and Andersen 2003), and where fees are charged per visit, the financial incentive to see too many patients jeopardises the quality of care (Chakraborty and Frick 2002).

Joint Public Private Initiatives are now promoted as a way of bringing private funding into health services. These include deals between governments and pharmaceutical companies over drug supplies. Problematically, they tend to be disease-selective; favour new technologies; are not always committed to sustainability and are limited to a fixed maximum level of resourcing (Save the Children 2001). Eflornithine, for example, nicknamed the 'resurrection drug' due to its effectiveness for treating children with African sleeping sickness, was provided at clinics run by Médecins sans Frontières between 1990 and 1998, but production ceased because it was commercially unviable. The manufacturer, Aventis, twice discovered and donated small amounts of eflornithine that had been overlooked, but refused to renew production. While providing the company with a philanthropic profile, such donations divert governments from longer-term objectives, and cause problems when funding ends (Save the Children 2001).

Most governments have been persuaded to undertake reforms in their health sectors. However, these have tended to be very complex and, it is argued, have diverted energies away from healthcare delivery (Simms *et al.* 2001).

Adolescent health

The health of young people aged 10 to 19, labelled 'adolescents' by WHO and other health professionals, has recently attracted international attention. Adolescents' health problems and needs, while different from those of young children, are also distinct from those of adults. Although generally the healthiest age group in any population, an estimated 1.7 million 10–19-year-olds die each year mainly from accidents, violence, pregnancy-related complications and illnesses that are preventable or treatable (WHO 2002a). Others develop chronic illnesses. Adolescence has a bearing on health in later life: malnutrition (estimated to affect 40 per cent of India's adolescents (Verma and Saraswathi 2002)) can cause permanent health problems, and it is estimated that 70 per cent of premature deaths among adults are attributable to behaviours initiated during adolescence (WHO 1998b). Furthermore, since many young people become parents during adolescence, adolescent health has intergenerational implications (WHO 2002a).

Adolescence is a time of rapid bodily growth and development, including sexual maturation, accompanied by changes in social relationships. Being a period of rapid change, adolescents include young people with widely different experiences and lifestyles:

> [a]t the lower end of the age range they consist of girls and boys, most of whom are not yet sexually active. At the upper end, they consist of physically mature young women and men, most of whom are sexually active and in many cases have children of their own.
>
> (WHO 1998b: 2)

Like other age groups, adolescents are differentiated by social divisions including gender, ethnicity, class and sexuality. Policy responses to adolescents' needs have nonetheless focused on a limited range of health issues: reproductive health (in relation to sexually transmitted infections (STIs) and pregnancy), mental health, substance misuse, injuries and violence. These reflect the social construction of young people as 'at risk'.

Reproductive health

Globally, reproductive health problems are the major cause of death for 15–19-year-old women (WHO 2002c). Girls who give birth before they are 18 are two to five times more likely to die in childbirth than women in their twenties (WHO 1998b). Those who have abortions are also at risk. Almost half the abortions carried out in sub-Saharan Africa are on adolescent girls (WHO 2002a): those performed illegally seriously endanger girls' health. Furthermore, each year, one in twenty adolescents worldwide contracts an STI (WHO 1998b), and the 15–24-year-old age group has the highest rate of new STIs (WHO 2002c). Every day, 7,000 young people become infected with HIV (WHO 2002b), with 15–24-year-old women five times more likely to become infected than their male peers (WHO 2002a).

International organisations have sought to explain the high levels
of unprotected sex among young people that appear to account for
teenage pregnancy and STIs in order to find ways to reduce them.
While earlier puberty, later marriage and diminishing influence of
family and culture are seen as underlying causes in many societies
(WHO/UNFPA/UNICEF 1997), much of the focus has been on
knowledge and access to services. Many adolescents receive little
information from schools or families about sexuality and
reproduction, sometimes because decision-makers believe that
withholding information deters young people from becoming
sexually active. In fact, research has shown that sex education
does not lead to earlier or increased sexual activity (WHO 1998b).
Young people also find it difficult to access family planning services,
including contraceptives, antenatal and obstetric care and treatment
for STIs (WHO 1998b) (Box 4.6). Research in Guinea found

Box 4.6

Reasons why adolescents are often excluded from health service provision

- Lack of knowledge on the part of the adolescent: young people may lack
  the knowledge to identify when they need to make use of health services,
  what services are available or how to access them.
- Legal or cultural restrictions: abortions may be illegal; health workers
  may be unwilling to supply condoms; young people may need parental
  consent for treatment.
- Physical or logistical restrictions: difficulties travelling to distant services;
  inconvenient opening hours.
- Poor quality of clinical services: poorly trained and unmotivated workers;
  lack of medicines and supplies.
- Unwelcoming services: young people are often very sensitive to privacy
  and confidentiality and likely to be deterred by long queues or
  bureaucratic procedures, or unfriendly or judgemental staff.
- High cost: young people may be unable to afford treatment, and are
  therefore dependent on parents or borrowing from other adults.
- Cultural barriers: young people may be inhibited about talking about their
  bodies or about sexual activity.
- Gender barriers: adolescent girls may be reluctant to be examined by
  men; young men may be unwilling to discuss intimate symptoms with
  female health workers.

Source: WHO (2002a: 21–22).

adolescents were five times more likely to visit a health centre for a pregnancy test than for contraception (WHO 2002a).

Married girls may have even less access to health knowledge and health services than their unmarried counterparts (WHO 2002a). In India, the median age at marriage is 16 for girls. Although girls do not necessarily join their husbands immediately after they marry, in rural areas 14.6 per cent are cohabiting by the age of 15 (Jejeebhoy 1998). They are soon under pressure to demonstrate their fertility: 36 per cent of married girls aged 13 to 16 are currently pregnant or already mothers. Contraceptive awareness is very low, only 59 per cent and 49 per cent of married adolescent women knowing of condoms and oral contraceptives respectively. Young married women have very little power in relation to sex or reproduction (Jejeebhoy 1998).

It is concern about HIV that has prompted most adolescent reproductive health interventions. Many governments and NGOs have pursued strategies to reduce HIV/AIDS infection rates among young people that focus on knowledge, attitudes and practices (KAP). It is believed necessary to provide youth with correct knowledge, and alter their attitudes in order to change risk behaviour. Social marketing of condoms, for instance, has been used as a way of making condoms acceptable – even attractive – to young people. New ways have been sought to present health knowledge to young people in such a way as to contribute to changing attitudes and behaviour. Peer education (including community or school-based AIDS clubs) and the media (Box 4.7) have, for instance, been used to both provide reliable information and encourage discussion and changing attitudes.

KAP approaches have had some successes – over a six-year period in the late 1990s the number of pregnant 15–19-year-olds in Lusaka infected with HIV halved (UNICEF 2001a). Most KAP interventions, however, have failed to inspire significant changes in sexual behaviour. Approaches that emphasise individual risk assessments and rational behavioural responses to knowledge are criticised for failing to take into account the social and economic contexts within which sexual relations take place, and even casting the blame for infection on young people who have no alternative than to engage in high-risk sex (Schoepf 2001). More structural approaches, such as that of Schoepf (2001) and Farmer (1992), stress the ways in which young people's agency is constrained by poverty and exploitative gender relations. For some young women, for example, sex is linked

Box 4.7

Adolescent health education in South Africa

Soul City – use of the media

Soul City is a joint venture of the South African government and NGOs which combines prime-time television, multi-language radio shows and easy-to-read booklets to provide accessible information on health issues. Health and development issues are integrated into a popular television drama series, now in its fifth season. The fourth series reached more than 16.2 million South Africans, two-thirds of them aged 16 to 24. The idea of the series is that, while entertaining, it encourages its audience to 'reflect on their own attitudes and behaviour, leaving them with a sense that they have a choice in determining their behaviour and the impact it has on their lives and those of others' (WHO 2002b: 23). The series has been shown to have increased accurate knowledge about HIV/AIDS and to have been influential in reducing resistance to condom use among young people. It has also stimulated discussion of social attitudes to violence, particularly against women and girls (WHO 2002b).

School-based peer education

Peer education 'involves the dissemination of health related information and condoms by members of target groups to their peers' (Campbell and Mac Phail 2002: 332). It is one of the most commonly used strategies for HIV prevention. It is based on the notion that sexual norms are socially negotiated within particular settings, and that the best way to encourage their renegotiation is to provide a setting for free discussion. Students are trained in participatory methods and given factual information about HIV and other STIs. However, the didactic culture of South African schools tends to undermine the concept. Peer educators tend to 'teach' rather than engaging in dialogue, and to focus on the biomedical 'facts' of AIDS, rather than addressing its social contexts, while teachers are unwilling to cede control over the process and seek to impose their own viewpoints. Ultimately, this form of peer education could reinforce young people's lack of power by creating situations in which they are instructed to engage in safe sex (or abstinence) without giving them insight into the social factors that make this so difficult (Campbell and MacPhail 2002).

to coercion and violence, and in many societies it is difficult for young women to refuse early marriage or to resist unprotected sex within or outside marriage, regardless of whether they are aware of the risks (WHO 1998b).

While it is helpful to remove the 'blame' for pregnancy and STIs from individuals, in so doing there is a risk of casting young people as powerless victims. In practice, the position may be more nuanced.

Plate 4.3 *AIDS posters, Lesotho.*

Source: Author.

Plate 4.4 *Drama performed by an anti-AIDS club, Malawi.*
Source: Author.

In many parts of Africa, in contexts of severe poverty, girls have sexual relationships with 'sugar daddies' – older men who are usually married. The power relations are clear: the young women exchange sex for material benefits, but are unable to negotiate details of the relationship, including, importantly, the use of condoms. Silberschmidt and Rasch (2001) argue that the girls are not simply victims, but social actors making deliberate choices. They do so, however, in the absence of adequate information: most are unaware of the risks they run, particularly in relation to HIV, but also to unsafe abortion should they become pregnant. Their use of their sexuality for material gain thereby makes them extremely vulnerable (Silberschmidt and Rasch 2001).

Failure to use contraception cannot always be explained by reference to unequal power relations. Gammeltoft (2002) undertook qualitative research in Vietnam, where, while young people increasingly engage in premarital sex, contraceptive use is extremely low. The young people explained their non-use of contraceptives in relation to the fact that sex was unpremeditated and governed by passion, not rational planning. Gammeltoft (2002) reads into the young people's accounts two key layers of explanation. First, in the current era of market economy in Vietnam, many people are concerned with being exploited by others. Young people need to demonstrate to sexual partners that they are motivated by pure feelings, and use of contraception might appear unduly calculating. Second, premarital sex remains socially unacceptable and morally illegitimate in Vietnamese society. Most young people subscribe to such a view, even if they themselves engage in sexual activity. By casting their own sexual experiences as one-off responses to sudden impulses, young people are able to retain a view of themselves as sexually innocent. Gammeltoft suggests that successful health promotion needs to address these larger social and moral systems rather than simply trying to change individual behaviour.

AIDS research and policy in Africa has strongly focused on prevention, to the neglect of treatment. 'Life skills are crucial, but may not be enough if a young person needs treatment that is not available' (WHO 2002b: 24). Very recently, attention has begun to turn to the needs of those already infected, and the fight for access to treatment has inspired activism among young people, especially in South Africa.

Mental health and suicide

Adolescents are widely perceived to be at risk from mental health disorders. At least four million adolescents attempt suicide every year and 100,000 succeed (WHO 1998b). Adolescent suicide rates are increasing in China and other parts of Asia, the Caribbean and Africa (WHO 1998b). Yet whether adolescence is universally a period of 'storm and stress' is highly contested (see Chapter 1). Margaret Mead's research in the 1920s suggested that although young people in Samoa experienced adolescence as a life phase, it was not particularly turbulent – indeed, young people enjoyed a period free from adult responsibilities. Others argue that there is a biological basis to adolescent stress: that hormonal changes are important, as well as changing relationships with parents and others (Schlegel and Barry 1991).

There are some psychological disorders (e.g. schizophrenia, bipolar) that have unquestioned biological roots and are prevalent among adults as well as adolescents. These generally become apparent during adolescence, and young people may have no frame of reference to enable them to understand what they are experiencing for the first time, or even recognise that they are ill or seek treatment, which makes them more vulnerable than adult sufferers (WHO 2002a). Other psychological disorders have been medicalised through the practice of adolescent psychiatry, but have seldom been identified among young people in non-Western societies (Fabrega and Miller 1995). Young people may experience some of the symptoms of these disorders, but they are not culturally understood as medical conditions. Anorexia, for instance, exists in some non-Western societies but is seldom associated with fear of fatness (Fabrega and Miller 1995). In many parts of Asia, 'possession' by supernatural agents is a common diagnosis of young people. While this is uncommon in the West, Multiple Personality Disorder is relatively common – but rare in Asia. Fabrega and Miller (1995) attribute these categorisations to cultural differences in understandings of the self: in Western societies where the self is seen as autonomous, dissociative disorders are marked by the presence of 'alternative selves'; in the non-West where the self is more socially defined, it is understood to be taken over from outside.

One case reveals the limitations of a biomedical approach in classifying psychiatric phenomena among non-Western people (Pineros et al. 1998). A group of nine young people from the Embera indigenous ethnic group in Colombia were afflicted with a condition that they attributed to a spell: the symptoms included fainting, headaches, convulsions and visions of people, animals or demons. Their community designated it as attacks of madness. Psychiatrists diagnosed a conversive disorder with dissociative features. The use of antipsychotic medicine, religious healers and traditional herbal remedies failed, but contact with shamans proved effective.

Smoking, alcohol and drug use

It is estimated that 20 per cent of smokers globally begin using tobacco before the age of 10 (United Nations 2002), and the vast majority begin by 19. Tobacco kills more people than any other drug. Worldwide, 300 million adolescents smoke: half of them will die of smoking-related causes in later life (WHO 1998b). Tobacco manufacturers, seeking new markets to compensate for declining sales in the health-conscious West, are increasingly targeting Third World countries, particularly their youth (WHO 1998b). Mortality from smoking related diseases is expected to rise to 10 million deaths a year by 2030 (WHO 2002a).

Use of other substances is also increasing. In Zimbabwe, the use by secondary school children of cannabis and inhalants has been shown to be greatest among the young people leading more Westernised and affluent lifestyles, although use of inhalants such as glue, petrol and aerosols also exceeds 10 per cent among both girls and boys in rural schools (Eide and Acuda 1997). Most policy relating to substance abuse has been geared towards individual remedial solutions rather than prevention. This is clearly inadequate given the scale of substance use and lack of resources in Third World cities. There is a need for policy to move away from focusing on individual psychopathology to consider wider social, economic and family pressures on young people (Boyden 1991).

Injuries and violence

Deaths and injuries from accidents are more common during adolescence than at any other age (WHO 2002a). Both injury and

---

Box 4.8

Gender and adolescent health

Most studies on adolescent health have focused on adolescent girls, yet boys, too, have health needs in adolescence. Boys are faced with a different set of health problems: '[b]oys world-wide show higher rates of mortality and morbidity from violence, accidents and suicide, while adolescent girls have higher rates of morbidity and mortality related to reproductive tract and pregnancy-related causes' (WHO 2000: 11). In most cases, boys suffer higher overall rates of morbidity and mortality in adolescence compared with girls. Eighty per cent of deaths among 15–24-year-olds in Sao Paulo, Brazil, in 1991–1993 were among young men. Men were found to be 20 times more likely than their female peers to die due to homicide (55 per cent of all male deaths in the age group), nearly five times more likely to die in a road accident, three times more likely to commit suicide and twice as likely as young women to die from other causes (Soares *et al.* 1998). Much social science research neglects gender-specific aspects of male behaviour and health, assuming that men are effectively genderless (WHO 2000). However, it is arguably male-specific behaviour that is responsible for the health problems faced by young men. Prevailing norms of masculinity in many societies are injurious to boys' health (WHO 2000). Boys commonly have difficulties seeking help and expressing emotions, which may have consequences for both physical and mental health. This is exacerbated if they spend a lot of time outside home and school settings and are therefore less well connected to informal and formal support networks (WHO 2000).

violence are highly gendered causes of mortality and morbidity (Box 4.8). In 2000, over 350,000 10–19-year-old men died as a result of unintentional injuries and violence (WHO 2002c). Road accidents are the single greatest cause of injury and death among 15–19-year-old men. In Africa, there was a three-fold increase in the number of deaths on the roads between 1968 and 1983. Many young people also experience accidents in the workplace. In many parts of the world, young men figure highly as both victims and perpetrators of violence (WHO 2002a). In major urban areas of Brazil and Colombia, violence is the principal cause of death among 14–17-year-olds (Pilotti 1999).

Conclusions

This chapter has focused on the most common areas of research and policy concerning young people's health in the Third World. The health of children under five years of age has received far greater attention than that of their older siblings, in large part because mortality rates among the under-fives are higher than among any other age group. Moreover, the diseases that commonly rob young children of their lives are diseases that have largely been eradicated in the West and are seen as particularly responsive to public health interventions. Studies of young children's health are dominated by a biomedical approach and by surveys that seek to establish causal mechanisms: while the behaviour of parents (particularly mothers) is studied, young children's voices are almost never heard in relation to their own health experiences.

The health of older children and adolescents is receiving increasing attention in both academic and policy circles. The concerns here are somewhat different: it is young people's individual health behaviours that are seen as problematic. Once again, however, the focus of research has been on risk factors, and altering the behaviour of individuals through efforts to provide knowledge and change attitudes. There is, however, a growing concern to listen to young people themselves and to recognise that they are social actors, who are sometimes constrained in their abilities to adopt healthy behaviours.

Key ideas

- While there are many alternative ways of thinking about health and illness, the Western biomedical approach has become dominant globally.
- International concern with the health of young people has focused predominantly on children under five, and more recently the reproductive health of adolescents.
- Child mortality (the proportion of children dying before the age of five) has fallen in most parts of the world, but 10 million children continue to die each year, mostly from preventable diseases.
- Although the very uneven patterns of child health are usually explained by reference to the behaviour of individuals and households, economic inequality and the actions of national governments and international organisations have much greater impacts.
- Although youth are usually the healthiest age group in any society, they do have particular health needs that are distinct from those of young children or older adults.

Discussion questions

- Assess the advantages and disadvantages of a biomedical approach to children's health.

- In what ways might conceptualising children in the Third World as social actors alter the nature of research conducted in relation to children's health?

- Should the immunisation of children against infectious diseases be compulsory in countries where such diseases are endemic?

- To what extent is it helpful to think of young people in the Third World as 'at risk' in relation to their health?

Further resources

Books

There are few books specifically relating to the health of children and young people in the Third World.

Boyden, J. and Gibbs, S. (1997) *Children of war: responses to psycho-social distress in Cambodia*, Geneva: UNRISD – explores children's mental health in a war situation.

de Waal, A. and Argenti, N. (eds) (2002) *Young Africa: realising the rights of children and youth*, Trenton, NJ: Africa World Press – has a couple of chapters relating to children's and young people's health in Africa.

Scheper-Hughes, N. and Sargent, C. (eds) (1998) *Small wars: the cultural politics of childhood*, Berkeley, CA: University of California Press – has a number of chapters relating to children's health, including provocative critiques of UNICEF's approach.

Journals

Social Science and Medicine frequently publishes accounts of research in this area.

Health and Place similarly publishes accounts of research concerning children's health.

Websites

IRC International Water and Sanitation Centre provides information about its School Sanitation and Hygiene Education Programme http://www2.irc.nl/sshe/.

UNAIDS http://www.unaids.org/ has an extensive range of online publications on topics including on mother-to-child transmission, children/orphans and young people.

UNICEF http://www.unicef.org/ is a useful source of information about children's health, and publishes online documents, as well as a catalogue of research publications produced by the Innocenti Research Centre http://www.unicef-icdc.org/.

The World Health Organisation http://www.who.int/en/ provides a considerable amount of information on health-related topics including adolescent health, child health and youth.

World Bank 'Multi-country reports by HNP indicators on socio-economic inequalities', http://www.worldbank.org/poverty/health/data/statusind.htm has useful cross-national statistical data on aspects of children's health.

5 Education

Key themes

- Forms of education
- Education for all? Patterns of educational expansion
- The benefits of education: competing perspectives
- The shape 'and quality' of education
- Reforming education.

Schooling is increasingly a feature of children's lives worldwide, and many remain in school into their late teens or twenties. The nature of education in a particular society depends in part upon how childhood is conceptualised in that society. Societies have very different ideas about how children should be expected to behave and what they should learn. The education delivered in different places is a product of culturally and historically situated beliefs and those promoted internationally.

This chapter begins with a brief consideration of some the world's diverse ways of educating young people. It proceeds to examine the global expansion of Western-style schooling, and the uneven patterns this has taken. The next section addresses competing perspectives on education: the perceived merits and drawbacks of educational expansion, along with critiques of particular forms of education. Children's experiences of even Western-style schooling vary from place to place, and some of the factors shaping individual experiences of education are considered. Last, the chapter examines some of the ways in which governments, often under international pressure, have sought to reform education.

Forms of education

All societies have ways through which young people acquire the skills and knowledge deemed important in adulthood. In the past, many children learned all they needed informally, in everyday interactions with their peers, older children and adults. Some societies, however, have long had formal education systems. Among the Basotho in southern Africa, informal learning was supplemented with short intensive periods of formal instruction where young people learned life skills and history, and undertook endurance tests. For over 700 years, children in Islamic societies across much of Asia and Africa have attended Qu'ranic schools or *Madrasahs*. Although these vary somewhat from place to place, they are generally single-teacher schools in which young people read, recite and memorise the Qu'ran, the primary purpose being religious rather than practical or economic.

The form of schooling that predominates globally today has its origins in Western Europe and was disseminated by Christian missionaries between the sixteenth and twentieth centuries. The missionaries' primary purpose was to enable people to read the Bible, but they also sought to spread what they saw as 'civilisation'. The education they introduced mirrored that provided to children of the lower classes in Europe. Colonial governments were often less enthusiastic about mass schooling, preferring to educate only a modest supply of low-level administrators (see Box 5.1). Nonetheless, the colonial era was one in which Western-style schooling took hold in most parts of the globe, and ultimately held within it the seeds of the end of colonialism. Most independence leaders of twentieth-century Africa and Asia were educated in Western-style schools, and attended universities in the West. Unsurprisingly, political independence often presaged a massive expansion in Western-style education.

Education for all? Patterns of educational expansion

Access to Western-style schooling has expanded dramatically, particularly since the mid-twentieth century. This is attributable to

Box 5.1

Education in colonial Rhodesia (now Zimbabwe)

The first mission school was founded at Inyati in 1883, but until British colonial rule began in 1890, missions had little success in teaching or evangelising (Dorsey 1975). With economic and political change in the 1890s, people embraced education. While the missionaries established schools to enable people to read the Bible and become evangelists, the colonial state wanted cheap pliable labour, and was unconvinced this required education. According to a member of the Rhodesian legislature in 1905, an 'uneducated native was the most honest, trustworthy and useful' (cited in Parker 1960: 72). Unable to prohibit missionary education, the state, from 1899, used grants-in-aid to control schools, requiring teaching of basic manual skills and diligence. The government took over urban schools in 1925 but missions remained responsible for rural education. The white nationalist Rhodesia Front, which came to power in 1962, sought further control over education. MP Andrew Skeen explained in a debate in 1969:

> We in the Rhodesia Front Government are determined to control the rate of African political advancement by controlling their education. Moreover, we wish to retain the power to retard their educational development to ensure that the government remains in responsible hands.
>
> (cited in Mungazi 1988:12)

Primary education was reduced from eight to seven years, black access to secondary education seriously restricted (Zvobgo 1986), and the churches forced to relinquish 2,308 schools, retaining only 635 (Zimbabwe 1982). Because the churches defended blacks against the government, and in particular their right to education, their ideas, however Eurocentric, gained widespread support. Education was a central demand of the independence movement and, in 1980, ZANU came to power on a manifesto promising free and compulsory primary and secondary education, and abolition of racial and sex discrimination (Dorsey 1989).

the enthusiasm of parents and children, national governments and international pressure. The right to free and compulsory elementary education is enshrined in the Universal Declaration of Human Rights and reiterated in the CRC. In 1990, participant countries at both the World Summit for Children and the World Conference on Education for All (EFA) at Jomtien, Thailand, pledged to work towards goals including universal primary education (UPE) by 2000 (Box 5.2).

While it is difficult to know exactly how many young people attend school (Table 5.1), the most recent major conference on education, the World Education Forum in Dakar, Senegal, in 2000, reported further progress towards UPE. The proportion of children in primary

Box 5.2

The six dimensions of the EFA targets

- Expansion of early childhood care and developmental activities, including family and community interventions, especially for poor, disadvantaged and disabled children;
- Universal access to, and completion of, primary (or whatever higher level of education is considered as 'basic') by the year 2000;
- Improvement of learning achievement such that an agreed percentage of an age cohort (e.g. 80 per cent of 14-year-olds) attains or surpasses a defined level of necessary learning achievement;
- Reduction of the adult illiteracy rate (the appropriate age-group to be determined in each country) to, say, one half of its 1990 level by the year 2000, with sufficient emphasis on female literacy to significantly reduce the current disparity between male and female illiteracy rates;
- Expansion of provision of basic education and training in other essential skills required by youth and adults, with programme effectiveness assessed in terms of behavioural change and impact on health, employment and productivity;
- Increased acquisition by individuals and families of the knowledge, skills and values required for better living and sound and sustainable development, made available through all education channels including the mass media, other forms of modern and traditional communication, and social action, with effectiveness assessed in terms of behavioural change.

Source: UNESCO (2000a: 13).

school rose over the 1990s from 80 per cent to 84 per cent and the absolute number of children not attending primary school fell (UNESCO 2000b). However:

> although there are more children going to school than at any time in history and more people can read or write than ever before, there are still 113 million children out of school, 97 per cent of them in the less developed regions and 60 per cent of them girls. While some regions, notably Latin America, the Caribbean and East Asia are on course to achieving universal access to primary education, other parts of the world are slipping behind. The problem is particularly marked in sub-Saharan Africa, with an increase in the number of children not in school.
>
> (UNESCO 2000b: 8)

Universal primary school completion is now one of the Millennium Development Goals to be achieved by 2015.

Table 5.1 *Measures of education*

Measure	Definition	Comments
Gross enrolment ratio (GER)	Number of children enrolled in school divided by number of school-aged children in the population	Easy to measure but often overestimates proportion receiving schooling because it includes: • Children enrolled in school who do not attend. (In Bolivia, teachers' pay is reduced unless all children in their communities are enrolled, but attendance is not enforced (Punch 2003).) • Children who are older than the official completion age, because they started late or repeated years. • Children younger than the official entry age, who are enrolled in school as a form of childcare. GERs often exceed 100% when education is rapidly expanding due to increased provision or fee reductions: children who had not previously enrolled in school (and may be well above school entry age) enter the first form of the cycle, and others who had dropped out return.
Net enrolment ratio	Number of children of school age enrolled divided by number of school-aged children	May underestimate school attendance by excluding children, particularly in rural areas, who begin school late, but may nonetheless complete the cycle. (In Lesotho, the primary GER exceeds 100% because of repetition, but the net enrolment ratio is below 60%. Adult literacy is 84%.)
Attendance rate	Number of children attending school regularly divided by number of school-aged children	Gives a more accurate picture of how many go to school, but data are seldom readily available.
Dropout rate	Children who leave school at a particular level divided by the number who began the level that year	Dropout is generally highest in the first year of each cycle.
Repetition	Children who repeat the level divided by those who began the level that year	This is most common where children have to pass end-of-year exams before progressing to the next class.
Adult literacy rate	The proportion of the adult population that can read and write a simple statement	It is extremely difficult to obtain internationally comparable school outcome data. Adult literacy is one of the more meaningful measures of the success of primary education. This is limited in scope and obviously incorporates lengthy time-lags.

Table 5.2 Education data for world regions

	Primary school enrolment ratio (%) 1995–99 (gross)		Primary school enrolment ratio (%) 1995–99 (net)		Net primary school attendance (%)		% of primary school entrants reaching grade 5 1995–99	Secondary school enrolment ratio 1995–99 (gross)		Adult literacy rate (%) 1990		2000	
	male	female	male	female	male	female		male	female	male	female	male	female
Sub-Saharan Africa	85	74	54	49	55	52	61	26	22	60	41	69	54
Middle East and North Africa	94	83	80	73	86	79	91	67	62	67	41	75	54
South Asia	99	81	79	66	74	68	59	57	40	60	32	66	40
East Asia and Pacific	106	107	92	93	95	95	93	65	61	88	72	93	80
Latin America and Caribbean	134	130	96	94	93	92	76	80	86	87	84	89	87
CEE/CIS and Baltic States	100	95	92	88	81	79	97	81	80	98	93	99	96
Industrialised countries	102	101	97	97	–	–	–	105	108	–	–	–	–
Developing countries	103	94	82	76	80	77	76	60	53	77	58	82	66
Least Developed Countries	88	74	62	57	57	52	62	31	26	53	31	61	40
World	103	95	83	78	81	77	77	66	61	82	69	85	74

Source: Based on UNICEF (2002c).

International concern has focused on primary schooling, partly because primary education is perceived to bring the widest social benefits at the lowest costs. Nonetheless, political pressure (not least from large numbers of primary school graduates who expect to proceed to secondary school) has resulted in the most rapid enrolment expansion being at secondary and especially tertiary levels. Enrolment is also uneven within each cycle. In many Third World countries, enrolment begins relatively high, but many children never proceed beyond the first year of school. The first year of each cycle is often swollen by the presence of children repeating the year. Dropout rates are often high, relatively few children remaining in school at the end of the cycle. In the Least Developed Countries, only 61 per cent of the children who begin primary school remain for the four years that is usually considered the minimum required to retain functional literacy.

Internationally, enrolment is very uneven (Table 5.2). Of the 113 million children not in school, 42 million live in sub-Saharan Africa (UNESCO 2000b), where due to high birth rates there were 24 million more children in 1998 compared to 1990 (UNESCO 2000b). There is a wide variation between countries within each world region

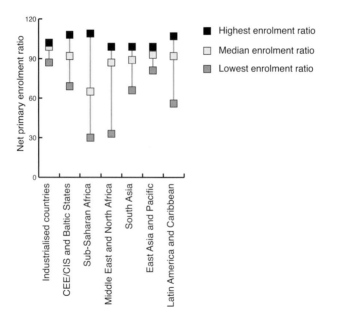

Figure 5.1 *Variation in primary school enrolment between countries within world regions.*

Source: Data from UNDP (2003).

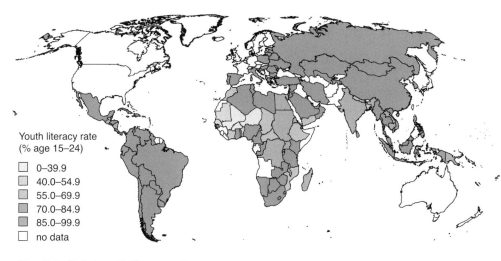

Youth literacy rate
(% age 15–24)

☐ 0–39.9
☐ 40.0–54.9
☐ 55.0–69.9
■ 70.0–84.9
■ 85.0–99.9
☐ no data

Map 5.1 *Global youth literacy rates.*
Source: Based on data from UNDP (2003).

(Figure 5.1). While only 12 per cent of primary-school-aged children attend school in Somalia, in neighbouring Kenya the figure is 74 per cent (UNICEF 2002c). In Zimbabwe, secondary enrolment in the mid-1990s exceeded 50 per cent, whereas in Rwanda it was only two per cent (Nsamenang 2002). Such disparities give rise to very different levels of literacy among young people (Map 5.1).

Within countries, discrepancies in school enrolment and attendance relate to both geography and children's gender, birth order, ethnicity, social class, religion and whether or not they have a disability. Urban areas are usually much better served with schools than rural areas (although the schools may be more overcrowded), as are regions with higher population densities, and more affluent residents.

The benefits of education: competing perspectives

Despite the global spread of Western-style education, this is not universally seen in positive terms. Formal education has a variety of impacts, the nature and extent of which are difficult to measure and widely disputed. These impacts are both immediate and in the future; affect individual students, their families (current and future) and society in general; and operate in various realms – economic, social, cultural, all of which are, of course, interrelated. Perspectives on the relative importance and desirability of education's impacts reflect differing conceptualisations of childhood, youth and the purpose of education.

Positive views of schooling

School attendance is widely believed to benefit individuals, facilitating cognitive development and providing valuable skills and knowledge for the future. Educated people are thought to have better economic prospects, better health and greater control over their lives. Education is also held to benefit society. If people can read and access information, democracy works better. Universal education in a meritocratic society is said to enhance social justice. Advocacy for increased access to education, however, is often premised on only a narrow range of these benefits.

The World Bank, IMF and many governments are inspired by human capital theory. Education is studied as an investment good, comparing expenditure on education (by the state, and by individuals in terms of fees, uniforms and earnings foregone) against other forms of investment. Profits (in terms of increased earnings for individuals or economic growth for the state) are assessed, and rates of return to education (ROREs) calculated. ROREs are demonstrably above the opportunity cost of capital (i.e. a worthwhile investment); higher in the Third World than the West; for primary education than other sectors; and higher for the individual student than for society (Psacharopoulos 1981).

There are many problems with this economistic approach to education, mainly relating to its neglect of the context in which schooling takes place. By focusing only on measurable outcomes, schooling is assumed to have no impact, for instance, on the informal sector or agriculture, despite their importance in many Third World economies, and non-economic impacts are disregarded. Power and politics are also neglected: there is no questioning of economic systems which accord different values to different people's labour, and hence affect how 'valuable' education is to them. Significantly, though, human capital theory has often been used to argue for improving girls' education (Box 5.3).

RORE findings are questioned by Dore (1976), who argues that education merely screens students by ability: those with most certificates are attractive to employers because they are the most able, not the best prepared for work. As more people obtain qualifications their 'value' diminishes. Students therefore compete

Box 5.3

Explaining and rectifying the gender gap in schooling

Sixty per cent of the children not enrolled in primary school are girls (UNESCO 2000b). Although there are 23 countries where more girls attend school than boys, there are 59 where boys' primary school attendance exceeds that of girls (UNICEF 2002c). Where the gender gap favours boys it is more often large and entrenched. There are 11 countries (Figure 5.2) where the gender gap in attendance is 10 percentage points or more (UNICEF 2002c).

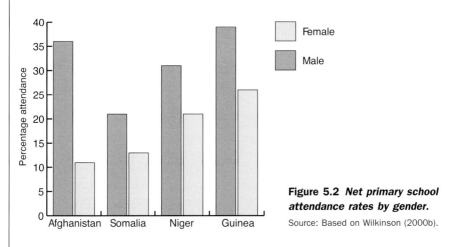

Figure 5.2 *Net primary school attendance rates by gender.*
Source: Based on Wilkinson (2000b).

The gender gap also tends to grow wider further up the educational ladder. Furthermore, even where girls attend school regularly, their academic performance may be worse than that of boys. There are several reasons why girls may be less likely than boys to attend or succeed in school (though these vary from place to place):

*   Girls are considered less likely than boys to need employment to support their parents or future families, and thus have less need for qualifications.
*   In some societies, it is assumed that girls are less academically able than boys.
*   Parents may fear for the safety and dignity of girls at school or travelling to school: in some societies girls are not permitted to mix with boys after puberty, and many schools lack toilet facilities.
*   The opportunity costs of girls' schooling may be greater than for boys: girls often perform crucial and valued work in the home and caring for younger siblings.
*   The direct costs are often higher: girls' uniforms may be more expensive and parents may feel they have to pay more to ensure that girls can travel to school in safety or live in secure accommodation when at school.
*   Girls may drop out of school to marry or because they become pregnant.
*   Schools may fail to interest or motivate girls.

In the past, girls' schooling excited little concern, but over recent years it has gained prominence. This is due to a growing recognition that the gender gap is damaging, not so much to girls themselves, but to society more widely. Girls' education is believed to bring greater economic benefit to Third World countries than that of boys. Educated women work in paid employment for longer than other women, increasing the tax base (Psacharopoulos and Tzannatos 1991); girls receive a higher wage premium than boys for each additional year of school (Schultz 1993a), partly because boys have access to highly paid, physically demanding work, irrespective of education (Deolalikar 1993); and while men respond to increased wages by working fewer hours, the opposite is true of women (Schultz 1993b). Girls' education is also perceived to contribute to reducing fertility and infant mortality, improving family health, and increasing the likelihood of children attending school (see Figure 5.2). It is these social benefits, conveyed by the proverb, 'when you educate a man, you educate an individual; when you educate a woman, you educate a family', that have inspired most policy on girls' education. Because returns to girls' education accrue more to society than to the individual, and because private costs incurred for girls' education may be higher than for boys', there is often less incentive for parents to educate daughters (King and Hill 1993). It is argued that society should therefore be willing to spend more on encouraging girls into school and reducing the costs for parents.

This very instrumentalist human capital approach neglects both the nature of the education girls receive and the impacts on girls themselves. Girls are not only less likely to attend school, but less likely to succeed in attaining the qualifications that education offers, for reasons relating both to their experience of school and constraints and expectations imposed from outside. There are barriers to girls' participation on an equal basis with boys: the attitudes of parents, teachers and other students who may assume that girls are less able, or have less need for qualifications than boys; school practices which do not allow or encourage girls to participate in all aspects of the curriculum (requiring them to learn 'domestic science' instead of physics for example); or expectations that girls should undertake chores such as sweeping or collecting water that remove them from lessons or reduce the time available for study. In school textbooks girls and women are often underrepresented or only depicted in sex-stereotyped roles. Providing girls with the same opportunities as boys, and presenting them with more positive role models, might enable girls to be more successful students.

Even increasing girls' academic success will not necessarily promote a more equal society, however. In Lesotho, female enrolment exceeds male at every level (Plate 5.1), and girls obtain more and higher qualifications than boys. However, whereas many boys find employment even without qualifications, girls usually require Cambridge Overseas School Certificate (COSC) to obtain paid work. Although girls are relatively successful academically, only a small minority pass COSC, and those who fail gain little from secondary education that is of use to them in a rural environment (Ansell 2002a, 2002b, 2004). This raises the question of whether most education systems are inherently androcentric: geared more to the needs and interests of boys than of girls. That said, in some societies, including the Anglophone Caribbean, concerns are growing about the failure of education to address boys' needs.

Plate 5.1 *Girls predominate in secondary school classrooms in Lesotho.*
Source: Author.

for ever higher qualifications with sufficient scarcity value to buy employment and this 'diploma disease' fuels enrolment. The problem is compounded in Third World countries with large income discrepancies, where certificates bring relatively greater rewards. While in Western Europe, university graduates earn on average two to three times as much as those with the minimum level of education, in some countries the ratio is 20 to 30 times. The difference in material contribution is not of this order (Colclough 1982), but graduate (and in some cases secondary school) recruitment operates in an international marketplace, and salaries must be competitive.

Critical perspectives on schooling

Most of the arguments favouring schooling focus on the advantages to society or to the young person in later life. Children tend to be seen more as 'human becomings' than social actors with their own immediate concerns. Education systems cast children as objects to be worked on by education, and children may come to see themselves in this way, regarding themselves as powerless to influence the school as an institution that would conform to their own wishes (Mayall 1994).

Since widespread schooling emerged in the West alongside industrial capitalism, it is understood by Marxist theorists not to empower individual students but to serve the needs of capitalism, by providing labour with necessary skills and dispositions. Schools are believed to operate as an 'Ideological State Apparatus' instilling in young people the dominant ideology, so that they accept prescribed roles in the economy (Althusser 1972). Many Third World education systems are said to provide young people with little that is of real value to them individually, but the disciplining of time and space is used to produce a compliant, disciplined workforce (Boyden 1991).

Equally, education may be interpreted as serving capitalism by stimulating consumption. According to Illich (1971), school serves to transform people's non-material needs into demands for commodities. In Kenya, for instance, 'luxuries' such as soap, tea and manufactured clothes were introduced along with waged employment. In time these luxuries became seen as basic needs and wage labour was needed to pay for them. To obtain wage labour, increasing levels of schooling were required, such that by 1977 education consumed 22.7 per cent of average Maragoli household income (Martin 1982). Illich asserts that no society can ever be rich enough to meet the demand for schooling, which constantly rises, costing disproportionately more at each level. The consumption level of a college graduate sets the standard for everyone else; expectations replace hopes, and failure to meet one's expectations gives rise to a sense of guilt. People are schooled into a sense of inferiority towards the better schooled and discouraged from taking control of their own learning. '[T]he poor are robbed of their self-respect by subscribing to a creed that grants salvation only through the school' (Illich 1971: 29).

However little value schooling has in its own right, lack of education, for example among girls, is used as a basis for discrimination. Far from eliminating inequality, the spread of education may increase it. Where most young people attend school, those who do not are disadvantaged economically and often socially. Even those who do receive education do not benefit equally from it. Schooling sets standards that some children cannot reach. Because school failure is often related to socio-economic background, schooling serves to transmit disadvantage from one generation to the next (Boyden 1991). Inequalities persist in society, even with equal access to education, and the poor (or girls) are blamed for their own failure (Ansell 2004). The wider power relations in which schools operate and young people study constrain education's capacity for social transformation.

Paulo Freire and critical pedagogy

Schooling has so far been represented as a monolithic and deterministic institution. Schools are in practice diverse and those attending them active social agents. Brazilian educator Paulo Freire (1972), offers a more nuanced critique of conventional schooling and its impacts on the poor (or in his terms 'oppressed'), advocating an alternative form of education that would enable the poor to participate in social transformation. Now labelled 'critical pedagogy', Freire's work has inspired many educators working in Third World contexts.

Freire's criticism of traditional education focuses on its concern with knowledge transfer. It is 'banking education' in which students engage in receiving, filing and storing knowledge, the best students being those who operate as meek receptacles. Banking education assists in 'cultural invasion' (Mayo 1995): the version of knowledge imparted is that which supports the dominant (generally Western) ideology and contributes to students' internalisation of their oppression. Students are immersed in a 'culture of silence' in which they are encouraged neither to question, nor to be creative.

Students do not necessarily act as passive receptors of knowledge, however: even in traditional schools, some students question what they are told. To encourage young people to resist oppression and transform society, Freire calls for a different type of education, focusing on consciousness-raising or 'conscientisation'. Students engaging simultaneously in academic and practical work would reflect upon the structural conditions underlying their physical labour, becoming aware that they are exploited.

Freire (1972) also criticises the opposition between teachers and students, in which the teacher takes the active position (teaches, knows, talks, disciplines) while students are passive recipients. He calls for power hierarchies to be broken down and students given an active role. By making student voices and experiences central, oppositional knowledges are recovered and education becomes more meaningful to students (Freire 1985).

Because formal education is part of the wider society, whose power relations it reflects and reproduces, Freire believed that before liberating education was possible in formal schools, society would have to be radically transformed. Instead, 'educational projects' carried out *with* the oppressed in non-formal settings might bring meaningful possibilities for change (Freire 1972). His ideas have been taken up by some NGOs, but have also influenced some formal education systems. However, it should be recognised that no intervention can take place in a context devoid of prevailing power relations.

The shape and 'quality' of education

Children's experiences of schooling differ greatly, for reasons associated with schools and education systems, and reasons associated with children's lives outside school. The shape taken by education in different contexts depends on a range of factors, among them historical legacies (particularly from colonial times), popular pressures, economic resources, political ideology and international influences. While development specialists are increasingly concerned about 'quality' of education, the impacts on young people are not one-dimensional, and the characteristics of, for instance, curricula and school organisation affect how young people see the world and shape their own identities in myriad ways.

Features of education systems

In countries that relatively recently emerged from colonial rule, education systems often continue to mimic those of the colonisers (even where European systems have themselves changed). Many former British colonies in Africa, Asia and the Caribbean continue to teach to examination syllabuses that are set and marked in the UK, as students' success in these 'gold standard' examinations (Cambridge School Certificate, O and A levels) is believed to lend international credibility to national education systems. Consequently, entire education systems at both primary and secondary level are geared to imparting the knowledge necessary to passing these exams, regardless of whether such knowledge is useful outside school. Although many countries are beginning to switch to more nationally relevant curricula and cheaper exams, children still follow academic and examination-oriented curricula (Ansell 2002b). Many also study using colonial languages, favouring the rich who may speak them at home (Boyden 1990). In Anglophone Africa, highest dropout rates often coincide with the level at which English-medium instruction is introduced (Bridges-Palmer 2002).

Teaching ratios and quality also vary between countries. Where education is rapidly expanding, it is seldom possible to train teachers quickly enough to keep pace with increasing student numbers. In such cases, student–teacher ratios are high and many teachers lack professional qualifications. Even where student numbers are stable, education budgets may be inadequate to train and pay sufficient qualified teachers to avoid very large class sizes and the use of unqualified staff. Furthermore, teaching is often among the worst paid and least respected public sector jobs, and poor working and living conditions do not encourage long-term commitment to posts, particularly in remote communities that lack electricity and running water. Unsurprisingly, teaching in poor countries is often subjected to intense criticism.

Apart from the criticisms levelled at teachers for their lack of professionalism (absenteeism, poor punctuality, failure to conform to local standards of morality), a more significant issue is the form of pedagogy that is often practised in schools in low-income countries. Education policy and teacher training have focused more on what is taught than on children's learning, with teachers being understood as technicians who convey knowledge from elsewhere rather than as reflective educators facilitating children's discovery of new knowledge (Dyer 2002). Prescriptive curricula and pressure to ensure that students pass public examinations further restrict teachers' creativity. Hence didactic teaching (Freire's 'banking' model) persists in many Third World schools (Shotton 2002).

Features of schools

No two schools are identical, but the extent to which the characteristics of individual schools are determined by the education system varies. Private schools differ from government schools, and vocational and non-formal schools (see next section) provide very different experiences. Even government schools differ considerably across a country and generally have some latitude to shape the education they provide, particularly at the preschool level (Box 5.4).

Plate 5.2 *A makeshift preschool outside a primary school, rural Mexico.*
Source: Author.

A key attribute of schools that affects students is accessibility. This depends both on location and the availability (and affordability) of transport (Gould 1993). In scattered rural communities, students may have to walk very long distances or there may be no school within reasonable proximity. The problem commonly becomes more acute at secondary level and above, particularly in places where few primary school graduates progress to secondary school. Boarding schools may be seen as acceptable (if expensive) alternatives at this level.

A school's teaching staff, buildings and resources generally reflect both national systems and local conditions. While teachers are usually paid by government, local communities may be required to provide other resources. Provision of teachers is usually prioritised. Lack of sufficient classrooms may mean students have to study outdoors or in a crowded room, perhaps sharing with other classes. In many rural areas, multi-grade classrooms are common, while some urban schools practise 'hot-desking' where some classes attend in the morning and others in the afternoon. Overcrowding and lack of classroom space is particularly characteristic of rapidly expanding enrolment. Ndirande Primary School in Blantyre, Malawi, for

Box 5.4

Preschools in India

One of the Education for All goals is 'the expansion of early childhood care and developmental activities' (UNESCO 2000a: 13). Many supporters of early childhood development and care (ECD), including the World Bank, draw on the American notion of developmentally appropriate practice (DAP), devised by neuroscientists and developmental psychologists, to argue that children in the Third World begin their education too late (often aged seven or more) and would benefit from structured learning at an earlier age (Penn 2002).

There is considerable debate as to the form such preschool education should take (if any). Many argue that children aged three to five are too young to participate in formal education: that they are not at the right stage of development to sit quietly and concentrate on a task, and being expected to may demotivate them in relation to education. While most Western childcare experts believe children learn best through play (through activities that are self-directed, exploratory and creative), others see this as rooted in Western ideals of childhood that stress individuality, autonomy, self-direction and self-esteem (Penn 2002; Prochner 2002).

Western play-based forms of early education were exported to India in the nineteenth and early twentieth centuries, but remained largely the preserve of an elite. Influential Indians have also favoured forms of early education focused on the development of 'the whole child'. Formal education below the age of seven was seen by Rabindranath Tagore as a corrupting influence, alienating children from their natural gift to learn and instilling conformity. Gandhi believed body and soul should be educated together emphasising the learning of crafts and spiritual training, rather than literacy (Prochner 2002). These ideas are little in evidence in India's preschools today.

Although 85 per cent of children entering Indian primary schools have no experience of preschools, 350,000 preschool centres that form part of the government's Integrated Child Development Services (ICDS) serve 10 million 3–6-year-olds (Prochner 2002). These centres have been very successful in health promotion, particularly through immunisation and feeding programmes, but the preschool education provided is of questionable quality. The centres lack resources and class sizes are very large. Furthermore, while those in charge of the curriculum and training advocate learning through play, parents prefer a more formal approach to literacy and numeracy. Ironically, the curricular flexibility permits the preschools to respond to parental demand, and in most classrooms children sit in rows learning to count and recite the alphabet, which parents see as the best preparation for the competitive atmosphere of primary schools (Prochner 2002).

There are also about 100,000 privately run 'preparatory' preschools in India. In these, play is considered fundamentally distinct from work, and therefore incompatible with learning. Instead the focus is almost entirely on reading and writing, usually in English, to enable children to compete for places in private primary schools. There are even pre-nursery schools which prepare children for nursery school admission tests (Prochner 2002).

The global trend for preschools to provide basic literacy and numeracy through rote learning reflects a felt need to prepare children early, to give them a competitive edge in a world of uncertainty, where most are likely to be confronted with poverty. The alternative international model based on DAP is no more interested in the welfare of individual children, being viewed primarily as a scientific means of improving 'human capital' (Penn 2002). Beyond these concerns with children's futures there are other reasons why donors and NGOs have eagerly promoted preschool education. One is that placing children in preschool centres of any form allows mothers to participate in the labour market. Another is the fact that preschools are commonly provided by the private sector at no public cost. In Brazil, there are even suggestions that private preschools have reduced demand for primary education (Penn 2002).

---

instance, has 21 classrooms and 9,000 students. Here, resources are scarce, enrolment high because education is free, and government-imposed building standards make classroom building expensive.

The quality of classrooms is not unimportant. Many students endure classrooms that are very hot, very cold, too dark to read or noisy when it rains, making it difficult to pay attention. Lack of electricity is a common feature of schools in the Third World, particularly in rural areas. This makes photocopying or the use of computers impossible, even if resources are available to buy such equipment (Box 5.5). Lack of water and sanitation can also be problematic: girls in particular are less likely to attend schools that have no toilets.

Other resources include furnishing (again dependent on finance – children, particularly in primary schools, may have to sit on the floor), and books and science equipment. In some places, especially where school populations are large, there are plentiful supplies of locally produced and relevant books, while other schools may depend on second-hand books imported from overseas, which may be both out of date and bear little connection to children's experiences. Access to science equipment depends on financial resources, often in combination with teachers' ingenuity.

While formal curricula are generally centrally determined, students also experience a school's 'hidden curriculum'. Most schools seek to ensure their students acquire particular behaviours and attitudes

Box 5.5

Schooling, imported knowledges and the digital divide

Education is seen by institutions such as the World Bank as a means of narrowing a perceived 'knowledge gap' between the West and the Third World (World Bank 1999). From a critical pedagogy perspective, however, Western knowledge is promoted in schools at the expense of local knowledges, a form of cultural imperialism that undermines students' self-respect.

In recent years, access to knowledge has been dramatically expanded through the Internet. The optimistic view is that this will 'democratise development', allowing people worldwide to both access and publish information. Young people are seen as early adopters of such technologies and various projects, such as DFID's Imfundo programme, have made Internet access available in Third World classrooms with the expectation of improving primary education. In practice, however, access to the Internet remains extremely uneven (see Map 5.2) and almost all the information available originates in Western countries. Fewer than one in a thousand Africans has access to the Internet and only 5.9 per cent of Internet hosts are located in the Third World (DFID 2000).

Young people do not passively absorb information, but negotiate the various, often conflicting, knowledges they encounter. Students in rural southern African secondary schools, for instance, mobilise two competing normative discourses when talking about gender relations. The idea of 'equal rights' is presented to them through both formal and informal curricula, while they also learn about their 'culture' through textbooks and curricula. They associate 'equal rights' with imported Western ideas, and therefore apply them more to those living Western lifestyles in town. 'Culture' on the other hand, represented to them as a static set of practices from the past, is applied by students to rural lifestyles. Although most students would like to live a modern lifestyle in town, in practice little employment is available and most girls in particular remain in the rural areas where they see traditional gender relations to be appropriate (Ansell 2001, 2002a).

through the day-to-day organisation of the school. These might be dispositions such as respect for authority or competitiveness, values such as fairness, or pride in local culture. Whether schools are hierarchical or democratic in their organisation impacts on children's experiences of school. 'Evidence indicates that schools involving children and introducing more democratic structures are likely to be more harmonious, have better staff/pupil relationships and a more effective learning environment' (Lansdown 2001b: 5). Participation in decision-making in school may also encourage children to see themselves as active agents in the wider world.

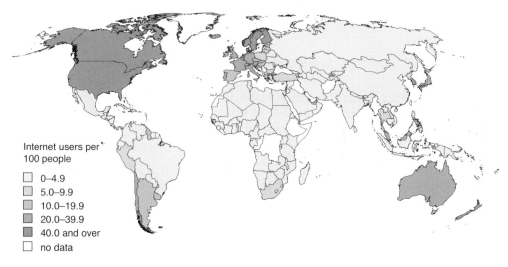

Internet users per
100 people

☐ 0–4.9
☐ 5.0–9.9
☐ 10.0–19.9
☐ 20.0–39.9
☐ 40.0 and over
☐ no data

Map 5.2 *Internet use, worldwide.*

Source: Based on data from UNDP (2003).

Plate 5.3 *Doing homework in Pakistan.*

Source: Lorraine van Blerk.

Children's wider lives

Children's experiences of education are shaped by wider events and processes in society (see Box 5.6), as well as their own characteristics and their families and communities. Parents' attitudes towards education, which may differ for children of different sexes, can have a strong impact. Some of the impacts of children's gender were addressed in Box 5.3. Both girls and boys often have to balance education with work. In rural areas, demands on children's time are often unevenly distributed throughout the calendar. In some countries, harvest is taken into account in the structure of the school calendar; in others it is not. Similarly, the length and organisation of the school day may be incompatible with young people's duties outside school. Many children are unable to do homework, as they have other demands on their non-school time, and may lack space or light to work. Boarding schools offer some young people welcome freedom from external pressures on their study time.

Box 5.6

Educating nomadic pastoralists in Mongolia

Nomadic pastoralists are seen as very difficult to reach with education, because they live in small communities and move around throughout the year. Unsurprisingly, millions of nomadic children remain outside formal education. Furthermore, where nomadic children and youth are provided with conventional education it may undermine their capacity to make positive and autonomous changes to their ways of life: schooling often damages pastoralists' self-confidence and frequently channels ambitious young people away from pastoral life and into the cities.

The situation in Mongolia has been rather different. This is due in part to exogenous factors – the way pastoralism is viewed by Mongolians and the political–economic organisation of the country – and in part to the way education was provided. About half Mongolia's population of 2.4 million live as nomads. Unlike in many countries, nomadic pastoralism is a respected way of life. Many who do not live nomadic lives have close kin who do, and many urban professionals entertain the possibility of adopting pastoralism. Mongolia is also unusual in that between 1950 and 1990 primary school enrolment among nomadic children increased from close to zero to almost 100 per cent, most attending boarding schools in rural settlements. After 1990, however, when Mongolia switched from a centrally planned communist state to a market-driven democratic republic, school enrolment among nomad children fell back to 75 per cent (Table 5.3).

Table 5.3 *Explaining changes in the school enrolment of nomad children in Mongolia*

Reasons for the communist government's success in enrolling young nomads in schools	*Reasons for the decline of school enrolment among nomad children in the 1990s*
Pastoralism was reorganised such that livestock became the property of collectives and were leased by families. A family could now herd livestock of a single species without exposing themselves to undue risk. This both created economies of scale, increasing the spending power of pastoralists, and reduced the demand for children's labour as the entire herd could be grazed together.	Abandonment of the collective ownership of livestock: nomadic families have increased their herd sizes and diversified their herds, creating a demand for children's labour.
School was completely free and teachers well paid and esteemed and therefore well motivated. This enhanced the respect for education among nomad families.	Education is no longer completely free, teachers are less well paid and less respected, and there is large-scale unemployment among educated people: schooling is no longer seen as so desirable a commodity.
The Mongolian culture revered nomadism. In spite of Soviet denigration of the practice, there was no sense within the education system that nomadism was an inferior way of life.	There is growing evidence that nomad children are bullied in school, being perceived as less able than urban children. This is exacerbated by a system whereby teachers are paid according to the success of their students: nomad children are considered less likely to succeed in school.
School attendance was compulsory and strictly and effectively enforced.	Laws governing school attendance are no longer strictly enforced.

Source: Kratli (2001).

Reforming education

Although parents, and to a much lesser extent students, exert some influence on education, it has remained the preserve of state policy to a greater extent than many other areas of children's lives. Education reform has tended to reflect national ideologies and agendas. Education is, however, a major public expense and, in countries that are indebted and subjected to structural adjustment, IFIs exert considerable influence over reforms to the sector, often seeking to shift responsibility for education away from government. Capital expenditure is often heavily funded by donors who negotiate their own preferred reforms.

Pro-poor education policies

For some years concerns have been expressed that education spending in Third World countries tends to focus on the upper end of schooling, which benefits an elite minority. In Africa, costs per student increase in the ratio 1:10:100 for primary, secondary and tertiary sectors (Bridges-Palmer 2002). Subsidies to secondary education tend to be around three times higher than those to primary education, while subsidies to tertiary education are up to 30 times higher (Addison and Rahman 2001). To redress this imbalance and move towards UPE, many countries are eliminating fees for primary education (Box 5.7).

Cost recovery schemes

While some countries are cutting school fees, others have recently reintroduced fees, to cut education budgets in accordance with SAPs and/or to raise funds to enhance the quality of schooling. The introduction of school fees has been shown to have adverse impacts on school enrolment. In Nigeria, primary education was made free and compulsory in 1976, and the historic gender imbalance was reversed in girls' favour. From the introduction of structural adjustment in 1986, however, girls' enrolment began to decline, more rapidly than that of boys (Obasi 1997). Girls, disabled children, the poor and those cared for by foster parents are particularly likely to lose out.

Another way of reducing the costs of education to the state is through the encouragement of private education. Private schools are held to be more efficient than the state sector, more responsive to parents' demands and to provide choice (Dyer 2002). In South Asia, minimally regulated private schools are mushrooming, teaching through the medium of English which is believed to enhance children's chances of securing better paid jobs. The poor, however, are unable to benefit from private education, and as private schools grow in popularity, the government schools that continue to educate the poor are increasingly viewed as inferior (Dyer 2002). Private education thus exacerbates socio-economic inequalities.

Decentralisation

Decentralisation is also promoted by those who wish to see a reduced role for central governments and increased local participation in

Box 5.7

Free primary education in Malawi

Malawi's first multi-party national elections in 1994 brought to power a government whose manifesto promised free primary education. There had been some cuts in fees prior to the election (non-repeating girls, for instance, paid no fees from 1992/93), but in 1994/95 when fees were eliminated enrolment expanded dramatically (Table 5.4).

The abolition of fees benefited poor and rural children disproportionately. Whereas in 1990/91 the primary GER among the richest quintile was almost double that for the poorest quintile, by 1997/98 the gap had narrowed and exceeded 100 per cent for all quintiles. The change in the distribution of net enrolment ratios was even more dramatic. The expansion of enrolment was also greatest in rural areas: while the urban GER grew from 115 to 119 per cent over the seven years, the rural GER increased from 77 per cent to 120 per cent. The gender gap, however, changed little. Secondary enrolment also increased considerably over the same period, again more markedly among the poor than among the rich.

Despite the improved enrolment figures, drop-out remained very high: in 1997 drop-out from standard 1 was 28 per cent. Costs, both direct (e.g. uniform, books, equipment, informal contributions) and indirect (e.g. loss of a child's labour or earnings), still deterred the poorest. There were also insufficient teachers: 18,000 untrained teachers were employed, but there were about 120 students for every qualified teacher, and teaching methods were often poor. Classrooms, teaching materials and sanitation were all inadequate. Many feel primary education has expanded at the expense of quality.

Source: Al-Samarrai and Zaman (2002).

decision-making, potentially enhancing relevance, quality, efficiency and stability (Dyer 2002). Local education committees are believed to have the interests of their own children at heart, and good knowledge of the needs and resources of their areas. Communities, however, are seldom easily defined, and are by no means homogeneous in their interests and concerns. Furthermore, decentralisation of education often involves a shift of responsibility for funding onto local communities, which are expected to provide buildings and other resources, and in some cases to pay for teachers. It can thereby exacerbate inequalities in education systems, rather than improving the situations of remote and marginalised communities.

Table 5.4 Changing primary gross and net enrolment in Malawi by gender and wealth

	Gross enrolment ratios						Net enrolment ratios					
	1990/91			1997/98			1990/91			1997/98		
	total	male	female	total	male	female	total	male	female	total	male	female
Poorest 20%	58	64	51	117	125	109	33	34	31	76	77	74
Richest 20%	110	113	106	120	129	113	75	76	75	80	80	81
Total population	81	86	75	120	128	113	51	52	50	77	76	78

Source: Adapted from Al-Samarrai and Zaman (2002).

Vocational education

Vocational education is often advocated as a response to the inability of labour markets to offer white-collar employment to all school leavers, and the widespread perception that conventional secondary education currently brings few benefits to the majority of students who are not successful academically. It is felt that education should prepare young people for work in sectors of the economy where demand for labour is higher.

Vocational education is not, however, homogeneous in its form or philosophy. Several broadly socialist African governments have attempted to integrate academic and manual work in schools, encouraging both reflection upon and respect for manual labour (as Freire recommended) and providing for the needs of young people with academic and practical leanings within the same institution. Education for Self-Reliance, a concept developed by Tanzanian president Julius Nyerere, aimed to invigorate rural life. Focusing on local practical skills 'it included an appreciation of local culture and customs, national identity and unity, cultural and moral values, preparation for both mental and manual work, and multilingualism including a local language' (Bridges-Palmer 2002: 99). Students and teachers would work together in production, students taking decision-making roles. Primary education was intended to be an adequate preparation for life, with little emphasis on examinations or progression to secondary school. Patrick van Rensburg developed the similar idea of Education with Production (EWP) in Botswana in the 1970s. Here, practical subjects were introduced into schools, and students joined self-help Brigades producing and selling goods and services to help cover the costs of their education. The emphasis of EWP was on community development through both education and production. In neither country was the integration of education and production fully implemented. Teachers who had been educated in an academic system, parents who expected academic education for their children and students who wanted academic qualifications did not value the time spent on agricultural or artisanal projects.

More recent attempts at vocationalising curricula have been inspired by a very different philosophy. Skills teaching is seen as a way of increasing human capital and contributing to economic growth. In Ghana, for instance, Junior Secondary Schools were introduced under the 1986 SAP. In addition to academic subjects, all students were required to learn a trade (Peil 1995). This vocational element, like EWP and Education for Self-Reliance, meets with resistance from those who aspire to work in white-collar employment:

> The government puts considerable emphasis on pre-vocational and vocational education. Its goal may be literate farmers and artisans who are satisfied with their lot and not longing for urban, formal sector employment. However, both parents and children see the matter differently. For them, development means a higher standard of living than agriculture or most forms of self-employment can provide, they are convinced that academic education is better than vocational education because it leads to higher income, [and] more prestigious jobs.
>
> (Peil 1995: 304)

Nonetheless, many young people do enter vocational training in Third World countries, usually having dropped out of the formal school system. Often such training is provided by the private sector and hence popular with IFIs.

Non-formal education

Non-formal education is seen as a way of reaching children who are difficult to provide with or attract into schools, including children in remote locations and those whose lifestyles are not easily accommodated to the routines of conventional schools (such as shepherds, street children and nomads). Non-formal education takes varied forms, from occasional meetings of a small group of young people with a teacher or facilitator, combined with workbooks to follow alone or with the group, to something closely resembling a formal school. In general, however, national curricula are abandoned or reduced, in favour of material deemed relevant to the learners. Instructors are usually not required to have formal teaching qualifications and may be drawn from the local community. Like vocational education, non-formal education is also more amenable to the involvement of NGOs. In Bangladesh, the Bangladesh Rural Advancement Committee (BRAC) runs 30,000 non-formal primary schools in areas where government provision is non-existent or insufficient (Dyer 2002). Release from the often rigid regulations imposed by education ministries on formal schools can allow non-formal schools to be flexible and innovative (Box 5.8), but the obverse is a lack of quality control or accountability. In Somalia, the only education is provided by NGOs, and often used to promote an

---

Box 5.8

Non-formal education in rural Mexico

In the early 1970s, Mexico established a National Council for Education Development (CONAFE) to serve rural communities where the small number of potential students did not justify the construction and staffing of a regular school. CONAFE trained young rural secondary school graduates from nearby towns to be elementary school teachers and equipped them with a 'Community Instructor's Manual'. The intention was to make the young teachers responsive to the communities, to want to stay permanently and to behave in keeping with the community's expectations.

In the 1990s, the scheme was extended to the postprimary level. Postprimary centres (Plate 5.4) were established in marginal rural communities, catering for students who had completed primary education. Although the majority of learners are young, adult learners also attend the centres. The key feature of the postprimary system is an open curriculum: students choose their own topics to study and engage in autonomous learning, using textbooks or other resources provided within or externally to the centre. Increasingly the centres are equipped with satellite television and Internet access. Because students learn at their own pace, those who cannot attend the centre daily are not disadvantaged.

The postprimary teachers or 'trainers' are educated in rural communities themselves, and have undergone secondary education and some college training in facilitating autonomous learning. They do not require a complete knowledge of the Mexican secondary curriculum, but assist students to learn for themselves. They work with a team of advisers at the regional level to put together a catalogue of 'didactic units' that students can work from, for which the trainer can provide both assistance and teaching materials, and which are geared to the needs and interests of the rural communities.

Source: Comara (1999).

Islamist agenda (Bridges-Palmer 2002). Furthermore, non-formal education projects are often short-lived and generally considered second best by parents and students.

Conclusions

While far more children in most parts of the world attend school than in the past, school attendance remains uneven. Education policy is most often considered in relation to what is best for society at large, rather than considering the lives of young people attending schools and colleges. Although children's voices are seldom heard in relation to education, schooling does have profound impacts on their

(a) Students presenting work.

(b) Science laboratory.

(c) Library.

Plate 5.4 *Non-formal postprimary education centre, rural Mexico.*

Source: Author.

immediate lives. It occupies a very large amount of time for most children worldwide, but also 'provides children with a space in which they can identify with the parameters of modern childhood. It makes possible negotiations with elders for better clothes and food; time for school, homework, and recreation; and often payment for domestic work' (Nieuwenhuys 1996: 244–245). Experience of schooling is highly influential in shaping young people's identities. Young people who have attended school often see themselves differently and have different expectations of their future lives compared with those who have not attended school (Ansell 2002a). 'The school is neither a miraculous medicine for all societal diseases nor an all-powerful poison' (Freitag 1996, cited in Nsamenang 2002: 90).

Key ideas

- All cultures have educated children, but not always in formal schools.
- Western-style schooling is expanding, but patterns of school attendance vary greatly between and within countries.
- Education is widely seen as desirable, but there are also powerful critiques of conventional schooling.
- Children's experiences of schooling differ depending on the characteristics of the education system, individual schools and children's own lives.
- There have been many efforts to reform schools, the most recent tending to shift control of education away from governments.

Discussion questions

- Should basic school education be considered a human right?
- Should governments concentrate on expanding the numbers of children attending school, or improving the quality of schooling?
- Consider the advantages and drawbacks of NGO involvement in the provision of education.
- To what extent should non-formal education be seen as a valuable alternative to formal schooling?

Further resources

Books

Freire, P. (1972) *Pedagogy of the oppressed*, Harmondsworth: Penguin – a classic text critiquing conventional schooling.

Freire, P. (1985) *The politics of education: culture, power and liberation*, Basingstoke: Macmillan – develops Freire's ideas further.

Gould, W.T.S. (1993) *People and education in the Third World*, New York: Longman – one of the few overviews of education in the Third World, although more from the planners' point of view than that of children.

Journals

Comparative Education.

Comparative Education Review.

Compare.

International Journal of Educational Development.

Websites

UNESCO http://www.unesco.org/education/index.shtml has information, statistics and online publications on a range of education topics, including considerable detail about Education for All.

UNICEF http://www.unicef.org/ is a useful source of information about education, and publishes online documents, as well as a catalogue of research publications produced by the Innocenti Research Centre http://www.unicef-icdc.org/.

The World Bank publishes statistics on education http://www1.worldbank.org/education/edstats/.

6 Work: exploiting children, empowering youth?

Key themes

- Categorising and enumerating children's work
- Where and why do children work?
- Should children work?
- Addressing child labour
- Youth employment and unemployment.

'Child labour' is often seen in the West as the unacceptable face of Third World poverty, yet in many societies it is considered entirely normal that children work. The problem of 'youth unemployment' is generally debated in very different circles, yet for many young people there is no sharp transition from child worker to employed youth, rather a continuity of experience. In practice, both discourses have roots in the modern Western notion that children should be in fulltime education until at least their late teens, then make a direct transition to fulltime work. This is not to deny that some children are adversely affected by the work they undertake or that some youth are adversely affected by lack of access to work. This chapter begins by considering children's work, and then moves on to address youth employment issues.

Categorising and enumerating children's work

Children across the world undertake a wide variety of work. A recent book about children's work in Zimbabwe (Bourdillon 2000b), for instance, discusses the activities of young street workers, vendors, domestic servants, carers, farm hands, tea and coffee pickers and miners: a far from comprehensive list. It is not simply the type of work that varies, but also the conditions under which it takes place. Children may work for families, themselves or employers; in the home, on the street or in a factory; for pay, for other remuneration or for nothing; a few hours a week or 14 hours a day (Table 6.1). Situations vary between children and from place to place. In Ghana, most child workers work part-time in family businesses and get some schooling, whereas in Pakistan many children work fulltime outside the family and do not attend school (Myers 1999). The immense variety of work, and wide age-range of the young people involved, complicates any discussion of causes and consequences, and conditions debates concerning the merits of allowing or discouraging children's work.

It is difficult to assess how many children work. Official economic activity statistics generally exclude children's work, particularly unpaid work, for several reasons: children are disregarded as workers; what they do may not be seen as work; and the often illicit

Table 6.1 *Categorising children's work*

Type of work	What does the work entail?
Degree of difficulty	Is it difficult, physically or otherwise?
Degree of hazard	Is the work hazardous?
Time commitment	How many hours? Daytime or night-time? Does it preclude attending school?
Place	In the home (the child's or another), on the street, in a factory?
Conditions	Is the working environment pleasant, comfortable, healthy and safe?
Labour relations	Does the child work for a member of their family, an employer or themselves? Are they working to fulfil a contract (e.g. bonded labour)?
Remuneration	Is there payment of any kind? Who receives the payment?
Reasons for working	Is the child working for their own benefit (for pocket money, to meet material needs, as a form of training) or because their family needs or wants them to work? Do they choose to work or are they compelled to do so?

Table 6.2 *ILO classification of children's work*

Children at work in economic activity		Children undertaking productive work in the formal or informal sector, including unpaid, casual and illegal work, but not domestic chores within children's own households
Child labour	in ages 5–11	All children at work in economic activity
	in ages 12–14	All children at work in economic activity minus those in light work (children may work up to 14 hours a week provided this does not harm their health and development or their education)
	in ages 15–17	All children in hazardous work and other worst forms of child labour
Hazardous work		Work in mining or construction, using heavy machinery or being exposed to hazardous substances or conditions; work of more than 43 hours per week
Unconditional worst forms of child labour		Trafficking; forced and bonded labour, including in armed conflict; prostitution and pornography; illicit activities including production and trafficking of drugs

Source: ILO (2002a).

status of children's work results in under-reporting (Boyden 1991; Robson 1996). Surveys designed to enumerate working children also encounter difficulties, not least in defining what constitutes work.

The International Labour Organisation (ILO), a United Nations agency, has estimated the extent of children's work worldwide. It employs a classification (Table 6.2) based on the place and conditions of work, degree of hazard, time commitment and child's age. The classification is not comprehensive: while the category 'children at work in economic activity' includes some forms of unpaid, casual and illegal employment normally omitted from official employment statistics, leaving out domestic chores excludes many children, particularly girls, who work fulltime in the home.

'Child labour' is the term the ILO applies to work that is deemed inappropriate because the workers are too young, or because it has adverse impacts on their well-being or education, or is considered hazardous. The 'unconditional worst forms of child labour', which involve approximately 0.5 per cent of children globally (ILO 2002a), are not considered in this chapter which deals with the forms of work that are most common. Prostitution and involvement in armed conflict will, however, be addressed in Chapter 7.

The number of working children globally is immense (Table 6.3). Although the proportion of children who work increases with age,

Table 6.3 *Numbers of children involved in ILO categories of work, by age and gender*

	All children	Economically active children		Child labour		Children in hazardous work	
	'000s	'000s	%	'000s	%	'000s	%
5–11	838,800	109,700	13.1	109,700	13.1	60,500	7.2
12–14	360,600	101,100	28.0	76,000	21.1	50,800	14.1
5–14	1,199,400	210,800	17.6	186,300	15.5	111,300	9.3
15–17	332,100	140,900	42.4	59,200	17.8	59,200	17.8
Boys	786,600	184,100	23.4	132,200	16.8	95,700	12.2
Girls	744,900	167,600	22.5	113,300	15.2	74,800	10.5
Total	1,531,500	351,700	23.0	245,500	16.0	170,500	11.1

Source: Based on ILO (2002a).

13.1 per cent of 5–11-year-olds are already economically active, more than half of them in hazardous occupations (ILO 2002a). The proportion of working boys slightly exceeds that of girls in every age group and category, particularly in the older age groups.

Of the world's economically active 5–14-year-olds, 98 per cent reside in poor countries (Table 6.4). In all poor world regions, over 10 per cent of 5–9-year-olds are economically active, reaching almost a quarter in sub-Saharan Africa (ILO 2002a). Among 10–14-year-olds the proportions in most regions double, with often a more modest increase into the later teens (Figure 6.1).

Where and why do children work?

There are three broadly distinct contexts in which children work: the family (either in the home or a family-run farm or business); the informal sector (for an employer or self-employed); or the formal sector (in, for example, a farm or factory). Differentiating between work contexts in this way allows some general differences to be explored, although the distinctions are by no means always clear-cut. Nor is it appropriate to assume that any one context is inherently better or worse than the others.

Within the family

Nieuwenhuys (1996: 245) argues that it is 'not so much their factory employment as their engagement in low-productivity and domestic

Table 6.4 Economically active children by world region

	All children 5–17	Economically active 5–9		Economically active 10–14		Economically active 15–17		Total economically active children	
	'000s	'000s	%	'000s	%	'000s	%	'000s	%
Developed economies	155,700	800	1.4	1,700	2.8	11,500	31.3	14,000	9.0
Transition economies	83,000	900	3.1	1,500	4.2	6,000	29.1	8,400	10.1
Asia and the Pacific	844,600	40,000	12.3	87,300	26.5	86,900	48.4	214,200	25.4
Latin America and Caribbean	139,300	5,800	10.6	11,600	21.5	10,300	35.0	27,700	19.9
Sub-Saharan Africa	207,200	20,900	23.6	27,100	34.7	18,100	44.8	66,100	31.9
Middle East and North Africa	111,600	4,800	10.8	8,600	19.6	7,500	31.8	20,900	18.7
Total	1,531,100	73,100	12.2	137,700	23.0	140,900	42.4	351,700	23.0

Source: Based on ILO (2002a).

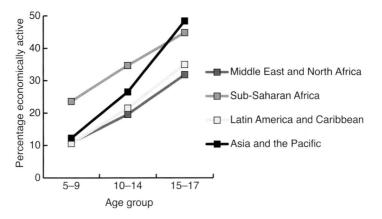

Figure 6.1 *Children's economic activity by age group.*
Source: Based on ILO (2002a).

tasks that defines the ubiquitous way poor children are exploited in today's developing world' (Box 6.1). Even very young children help in peasant farming households, their tasks increasing as they grow older. In northern Nigeria, children's unpaid work for the family includes childminding, household chores, tending the fields, feeding livestock, serving in shops, working in home industries, street vending, running errands and fetching water (Robson 1996). These tasks are gendered and differentiated by age. Although there are considerable differences within and between households, the time commitment can be considerable (Table 6.5). By 10 to 12 years of age some children contribute as much as adults to household sustenance (Robson 1996).

Work for the family may demand a high level of responsibility. In rural Lesotho, two-year-old boys begin herding livestock with older siblings and friends, and by six or seven many are herding fulltime (Kimane and Mturi 2001). Charged with the household's wealth in the form of cattle, some boys stay away from home for weeks at a time sleeping at makeshift cattle posts on remote mountainsides, many encountering armed cattle thieves and severe weather.

The ILO distinguishes between 'productive' work within the family, and 'domestic chores' which are not regarded as economic activity. Productive work is that which generates produce for subsistence or sale, such as agriculture or handicrafts. Domestic tasks such as cooking, cleaning, fetching water and childminding contribute to household welfare but do not generate a specific product. The

Table 6.5 *Domestic work undertaken by rural Hausa children in northern Nigeria*

	Boys			Girls		
	% reporting domestic work	Average time spent (hrs/day)	Range time spent (hrs/day)	% reporting domestic work	Average time spent (hrs/day)	Range time spent (hrs/day)
Aged 6–9	89	2.1	0.8–5.0	100	1.9	0.3–4.8
Aged 10–15	77	2.1	0.5–5.5	100	2.5	0.7–6.3
Total	81	2.1	0.5–5.5	100	2.1	0.3–6.3

Source: Based on Robson (2004); children also engage in productive and trading work.

Box 6.1

Working children in rural Kerala

In Poomkara, part of a large coastal village in Kerala, India, girls aged 5 to 15 spend an average of six hours a day on domestic work and about an hour and a half on productive activities such as spinning coir yarn; among boys, domestic work demands about four hours a day and a further three hours are devoted to productive work, mainly in the fishing industry. Most such work takes place within the family, although some children are hired out to neighbours to help in their businesses.

Even unpaid family work is related to wider social and economic processes. Both fishing and coir manufacture in Kerala depend on access to unpaid or very low paid labour. As the fishing industry has expanded in response to growth in foreign markets since the 1960s, the demand for boys' work has increased. In contrast, overseas demand for coir yarn peaked in the 1950s and has since steadily declined. Rather than releasing girls from their work, however, coir manufacture now depends more on unpaid workers. Girls, nonetheless, work shorter hours than in the past and, owing to the Kerala government's efforts to promote girls' education, most combine work with school.

Despite the importance of children's unpaid work to both individual households and the Kerala economy, children's tasks are valued little by local people and may not even be regarded as work. Girls' work, in particular, is seen simply as devotion to the family – a pattern that continues into adulthood. Girls are expected to put family interests above their own, and cannot retain any money they earn. Boys, by contrast, may be allowed to spend some of their earnings on themselves.

Source: Nieuwenhuys (1994).

(b) Feeding chickens, Lesotho.
Source: Rebecca Kenneison.

(c) Combining childcare with farm work, Nepal.
Source: Author.

(d) Herding in the snow, Lesotho.
Source: Author.

(a) Making bread, Lesotho.
Source: Author.

Plate 6.1 Working for the family.

distinction is rather arbitrary: when undertaken for payment, domestic tasks are classified as 'productive'. Arguably, the distinguishing feature of domestic chores is that they are usually performed by women or children: thus rendered 'not real work', domestic chores are among the least valued of children's work.

Domestic work is often time consuming and can be very arduous. It may involve carrying heavy loads of water or fuelwood. Even in urban households, girls may be expected to get up before everyone else and prepare breakfast. Some children undertake particularly difficult domestic work. Those who care for sick or elderly relatives, for instance, may be confined to the home for long periods and perform work that is not only physically taxing but also emotionally challenging (Robson 2000).

The family is not always clearly demarcated. Children's work often plays a role in maintaining kinship relations: children are sent to work for wealthier kin as domestics or apprentices, sometimes to fulfil an obligation. Although condemned internationally as 'bonded labour', work pledged between Indian kin sometimes operates as a valuable form of training (Nieuwenhuys 1996). In Zimbabwe, some children are fostered by distant relatives and given kinship titles, but although they appear to outsiders as family members, they may be expected to work much harder than closely related children (Bourdillon 2000c).

Informal sector

Most types of work performed by children within the family may also be undertaken for pay, outside the family. Children are employed by households where there are no children (or insufficient children of the right age or gender) or where the household's own children are expected to spend their time in education, leisure or even other forms of work. They help with family businesses, piecework, farming, herding or domestic work (see Box 6.2). Children also work as vendors, producers and small-scale service providers in the informal sector, either for employers or for themselves. Activities include newspaper selling, shoe shining, collecting minibus fares, hawking, portering, rag picking, fishing, guarding/washing cars and handicrafts. Informal employment can be hazardous. Children who break bricks in Bangladesh fear crushing their fingers under a hammer or being blinded by a flying chip of brick (Woodhead 1999). Boys collecting fares often cling to the

(a) Shoe-shiners, Delhi, India.

Source: Joy Ansell.

(b) Market trader, Pakistan.

Source: Lorraine van Blerk.

Plate 6.2 Work in the informal sector.

outsides of minibuses to call out to potential passengers: every year some fall and are killed (Woodhead 1999). Most types of work have advantages and disadvantages. Research in the Philippines found most child workers evaluated their own employment positively in relation to alternative jobs (Woodhead 1999). Within the informal sector, children's work is generally unregulated, although the boundary between formal and informal sector work is

---

Box 6.2

Child domestic workers

Throughout the world, children work as childminders, maids, cooks, cleaners, gardeners and general domestic helpers in other people's homes (Pflug 2002). Unlike much informal sector work that takes place in the street, domestic work happens invisibly within the private sphere. Labour force surveys often fail to enumerate children employed as domestic helpers: many are unpaid, and those working for relatives may be assessed to be family members. This makes it difficult to gauge the exact numbers involved.

In Dhaka, Bangladesh, an estimated 250,000–300,000 girls work in domestic service (Seabrook 2001). Their ages peak at 10 to 11 and again at 14 to 15 (pre-pubescent girls are sometimes thought less sexually attractive to the household's men). Although most receive US$ 2–4 a month, some are simply fed and housed. Domestic work has a low status, but is believed to provide the potential for social mobility. Many parents see the shelter of a middle-class household as a safe environment that might improve their daughter's marriage prospects. In practice, however, many girls are abused, both physically and sexually. Domestic work also requires girls to negotiate new identities in contexts where little value is attached to their work, and where social customs may differ from their own. Furthermore, with no contracts or fixed limits to their duties, they tend to do whatever they are asked.

In Metro Manila, the Philippines, about 30,000 children work as domestic helpers (Camacho 1999). Some, as young as seven, wash and clean for neighbours. Many older girls, in consultation with their families, decide to move to the city to work as domestic servants. Some earn wages and may send money home, but anyway their absence from home reduces the number of mouths to feed. Most also spend money on personal items and sometimes school fees. Other child domestic workers are unwaged but work for food, clothes and the opportunity to attend school. Many have no time for education, however, working daily from 6a.m. to 10p.m. The involvement of family-based social networks in recruiting domestic workers limits serious abuse. Unlike in Bangladesh, some child domestic workers benefit from social security registration, but their work has low status and many employers are reluctant to respect their rights as workers.

not entirely clear-cut and in some countries young people employed in small-scale industrial and service sectors benefit from written contracts and labour legislation.

Formal sector

Work in export-oriented factories and farms has dominated Western concerns about child labour, yet while the numbers are not insignificant (800,000 young women in Dhaka work in garment factories), the formal sector employs only five per cent of working children (Seabrook 2001). Conditions in factories can be exploitative. In 1986, eighteen 12–14-year-old girls were rescued from a textile mill where they were locked in, beaten, and worked from 5a.m. to 9p.m. daily (Boyden 1991). Children in Indian match factories are exposed to noxious fumes for 12 hours a day and face risks of fire and explosions (Boyden 1991). Children gluing shoes in Brazilian factories have become glue addicts (Scheper-Hughes and Hoffman 1998). Terms of employment may be strict, for instance denying children time away from work. Yet in many cases the formal sector offers better conditions and higher pay than the informal sector. In Bangladesh, many girls in informal sector employment aspire to work in garment factories (Woodhead 1999).

Even in a single industry, conditions vary. In Nepal, rural Tamang and Thamai families send children to work in carpet factories. Research found girls from a village supplying workers to one factory aspired to factory work, employees reporting the work was easier than at home. By contrast, children from a village that supplied a different factory unanimously said they would rather remain in the village, because the factory required continuous work from early morning until late at night, with little time for play. They also complained about factory conditions and the prevalence of disease among the workers (Johnson *et al.* 1995).

Why children work

The reasons why children work vary within and between societies, but generally relate to cultural and structural factors as well as the agency of children and their families. In many societies, children have always worked, family survival depending on the roles performed by children, in both productive and domestic spheres.

It is commonly asserted that most child labour is attributable to poverty. Worldwide, most working children live in poor countries and are drawn disproportionately from poorer households. Some economists believe that credit markets are to blame: if the poor could borrow to invest in their children's education for future benefit, they would not have to send them to work to meet immediate needs (Brown 2001). Others, however, argue that education holds little advantage for the poor in low-technology societies. Children from the poorest third of Indian society often join the workforce early because the education system does not appear to lead to employment (Verma and Saraswathi 2002). If children's productivity is similar to that of adults, the use of child labour is economically rational.

A contrasting view is that child labour is explained less by poverty than by labour-market conditions. In urban Brazil, for instance, employment rates among 14–16-year-olds increase as labour-market opportunities improve (Duryea and Arends-Kuenning 2003). Other explanations rest on the growing culture of consumerism that creates demand for income and propels children into employment to facilitate the acquisition of consumer goods both by children and their families.

Whether the motivation is poverty or material acquisition, the decision to work is usually taken by children or their parents. The extent of children's choice of whether and where to work differs between societies. It is often difficult for children to refuse to work, if this is what their families want, especially where the family depends upon it. Other than in situations of forced labour, however, it is rare for children passively to do as they are told, and not take any part in decisions regarding the work they will perform (Bourdillon 2000c). In some cases, children exercise considerable agency (Camacho 1999).

Should children work?

Many trade unions and the ILO are fundamentally opposed to the employment of children. Others see nothing inherently wrong with children working, and even defend children's right to work. For most children, work entails costs and benefits, and the acceptability of any particular work can be assessed in relation to the balance between these (McKechnie and Hobbs 2002).

Arguments against children's work

Article 32 of the CRC binds signatories to recognise:

> the right of the child to be protected from economic exploitation and
> from performing any work that is likely to be hazardous or to interfere
> with the child's education, or to be harmful to the child's health or
> physical, mental, spiritual, moral or social development.
>
> (United Nations 1989)

Opposition to children's work usually centres on potential hazards
and the effects on children's bodies. Children are physically more
vulnerable than adults to some hazards. They 'are more liable than
adults to suffer occupational injuries, owing to inattention, fatigue,
poor judgement and insufficient knowledge of work processes and
also because the equipment, machinery, tools and layout of most
workplaces are designed for adults' (Bequele and Boyden 1988: 3).
Research has established that working can have adverse
consequences for children's growth (Ambadekar *et al.* 1999), and
for their future health (Kassouf *et al.* 2001). The level of harm
relates to the type of work and conditions under which it is
performed. Carrying heavy loads or long periods of close work can
damage growing bodies, and exposure to chemicals can carry greater
risks for those who are sexually immature (Gailey 1999).

A second central concern is the impact on children's schooling.
Clearly children in fulltime work are unlikely to attend school
(although some do combine eight hours of work with a half-day shift
at school). Many poor Paraguayan households keep one or more
children out of school to work: out-of-school 12-year-olds in formal
employment contribute about a quarter of household income
(Patrinos and Psacharopoulos 1995). Failure of poor children to
attend school can perpetuate poverty and exclusion. Even where
children combine work and school, they may be less successful
educationally than their non-working peers. In Ghana, working
children do worse in school, even where their attendance is
comparable with children who do not work, perhaps because of
exhaustion or because their interests now lie elsewhere (Heady
2000).

Other opponents of children's work make accusations of exploitation.
Some work that children undertake is very hard and poorly paid
(if paid at all), and children themselves often receive no
remuneration. An 11-year-old in Zimbabwe was found to be working

12 hours a day, 7 days a week on a farm, for which his parents were paid about $3 a month, one eighth of the minimum agricultural wage at the time (Bourdillon 2000c). Similar cases, while by no means typical of children's work, are not infrequent. Children are useful to employers as they can be laid off easily when business is slack, are cheaper than adults, have no rights as workers and cannot join unions (Bequele and Boyden 1988). Although there may be practical reasons for employing children (small fingers may be advantageous in some occupations), they are also, significantly, willing to undertake work that adults consider degrading or unpleasant (Boyden 1991). People also use children to do illegal work to avoid being arrested themselves. Trade unions believe children's workforce participation negatively affects adult employment. Adults' bargaining power is diminished where they could be replaced by children who can be persuaded to undertake the same work for less pay and in worse conditions (Myrstad 1999).

It is not only employers that are held to exploit children's work. In situations of poverty where most children work, children may be seen by their families in instrumental terms, valued for their economic contribution rather than for themselves (Bourdillon 2000c). Some demographers believe the persistence of child labour encourages high fertility rates, as people have many children in order to benefit from their work (Dessy 2000). While these views are contested (Lieten 2000), those concerned with curbing population growth tend to favour the elimination of children's work.

Not all work undertaken by children is hazardous, exploitative or damaging to education, yet some people simply believe that it is inappropriate for children to work: that working deprives children of their childhoods, which should be spent on leisure and education (Box 6.3). Certainly working reduces children's free time, and may reduce time spent with friends and family, but the importance attached to leisure in childhood is culturally determined.

Arguments in favour

Arguments supporting children's involvement in work reject the Eurocentric notion that work is not an appropriate part of childhood (Box 6.3): in most societies it has always been considered normal that children work. Furthermore, protagonists assert children's right to work, citing Article 12 of the CRC: the right to express an opinion. When consulted, most working children emphasise their desire to continue to work, although some argue for better pay and conditions.

Children may benefit in a number of ways from working. Many take pride in their contribution to their families' welfare. Knowing that their earnings are spent on rent, food or a sibling's school fees gives many Bangladeshi children a sense of responsibility, or even indispensability, within the family (Seabrook 2001). Working for

---

Box 6.3

'Work-free childhood'

Children's work has been condemned in some circles in the West since the early-nineteenth century. Today most working children live in the Third World, but the debate about child work remains dominated by organisations based in the West (Seabrook 2001). Whereas the nineteenth-century movement against child labour was founded in moral outrage at Victorian factory conditions, the contemporary focus is the infringement of children's right to a labour-free childhood (Seabrook 2001). The view that children should spend their time at school and play is a relatively recent middle-class perspective, but has become dominant among international NGOs and experts (Myers 1999).

The belief that childhood should be work-free arguably relates less to concern about work itself than about children's involvement in the economy. After all, idleness in children is usually perceived in negative terms and children are expected to work hard in school (Bourdillon 2000c). Children's work is seldom considered problematic provided it remains outside the monetary economy: a situation that reflects both romantic notions that children's lives should not be tainted by financial considerations and fears about children gaining economic independence.

While many NGOs and international organisations share a 'zealous belief in the desirability of extending Western childhood ideals to poor families worldwide' (Nieuwenhuys 1996: 241), children do not inhabit a blissful land of play set apart from politics and economics, but are affected by and have to participate in a real world of production and consumption (Bourdillon 2000c). The ILO describes child labour as 'work that deprives children of their childhoods' (Martin and Tajgman 2002: 16), yet stigmatising working children in this way does not help those with no alternative (Woodhead 1999). Excluding children from the production of value may reinforce their vulnerability to exploitation (Nieuwenhuys 1996). Nevertheless, 'countries that are affluent enough to prohibit their children from earning money, portray this ability as moral superiority by condemning, as inhumane, those who encourage children to do productive work' (Bourdillon 2000c: 7). The ideal of a work-free childhood is now infiltrating countries where many children work, leading to a spectrum of views on childhood (Bourdillon 2000c).

themselves makes children more autonomous and self-reliant and may also give them a sense of self-respect (Bourdillon 2000c). Furthermore, children learn through working. Yoruba children in Nigeria who are sent on errands and perform domestic duties demonstrate greater cognitive development than their non-working peers (Nsamenang 2002). Working children may develop specific skills, or simply become accustomed to working practices, and socialised into adult life, which may make them more adaptable and able to participate in the workforce more effectively in the future. Many African parents believe hard work makes children more resilient as adults (Rwezaura 1998).

Children may be economically empowered through work, enabled to meet their own needs, or at least bargain within the family. Even if adult wages were higher in the absence of child labour, there is no guarantee that adult earnings would be spent improving children's lives, yet '[t]he notion that children's economic freedom should be abridged to protect the economic welfare of adults has long been a pillar of trade union doctrine and much social policy' (Myers 1999: 15).

Those who favour children's right to work contest the way work is considered an obstacle to education. Some children work to earn money for school fees and uniforms. There is also little conclusive evidence that increasing school enrolment removes children from the labour market. The expansion of schooling in Kerala did not reduce children's work, but added to their duties (Nieuwenhuys 1994). It may therefore be more appropriate to adapt the school system to make it compatible with work than to expect children to stop working to attend school (Admassie 2003). Internationally, attitudes towards work contrast with those towards education. Poor-quality schooling that holds little interest or use for children receives much less criticism than paid employment (Bourdillon 2000c). While the CRC grants children a right to be *protected* from work, school is compulsory, even though children may evaluate it in similar terms to work (Woodhead 1999). Children's right to earn money has few defenders (Miljeteig 1999).

Acceptable and unacceptable work

A conceptual distinction between 'children's work' which is considered acceptable, and 'child labour' which is condemned, is gaining favour, including with governments of Third World countries that see it as more realistic than outright opposition to children working (Myers 1999). Where and how the boundaries are drawn is, however, seldom clear and subject to considerable contestation. Any assessment of whether a particular child's work is acceptable must relate to the nature of the work, the conditions under which it is performed and the terms of employment (Miljeteig 1999), as well as the age and abilities of the child and whether or not they are also expected to attend school (Bourdillon 2000c), but the judgement is also necessarily subjective and culturally situated.

A key issue is whether children themselves benefit from working. Work that improves children's future chances, even if it is unpaid, is likely to be looked upon favourably (Gailey 1999). Apprenticeships and training are generally regarded as acceptable, as children are expected to acquire a valuable skill. Apprenticeships can, however, be guises for the employment of cheap labour. Other forms of work have also remained acceptable in the West, long since children were removed from factories, because they are deemed to provide opportunities for socialisation and training. These include housework, childminding, helping on family farms and in small shops, running errands, newspaper rounds and seasonal farm work (Nieuwenhuys 1996).

The acceptability of much of this work relates both to the absence of immediate returns to the children involved (there is seldom significant payment involved) and also the context in which the work takes place. It is commonly supposed that children's waged work is qualitatively different from work traditionally undertaken in domestic settings, as it involves an employer–employee relationship in which children are considered more vulnerable than adults (Bequele and Boyden 1988). Nieuwenhuys (1996), however, argues that children who work unpaid are exploited as much as those in factories. Seabrook (2001) contrasts the 'child labour' of boys in Bangladeshi cigarette factories with the 'children's work' of their sisters who roll cigarettes at home for one-third of the boys' pay. Even domestic work in the home can remove children from school, and work on a family farm can expose children to chemicals and other hazards (Bourdillon 2000c). However, the ideology attached to the family makes it harder to condemn children's work within family contexts.

Another consideration is whether children work by choice or coercion. Johnson *et al.* (1995) argue that where children feel guilty about not carrying out household tasks, domestic work becomes 'child labour'. Yet the sense of responsibility associated with having to work may be valued. The reason for work is also significant. Bourdillon (2000c) cites the situation of a child required to stay home from school for urgent work on the fields. If there is a need to bring in the harvest before approaching rain to prevent it rotting and leaving the family short of food, Bourdillon argues the work is acceptable; on the other hand, if the father wants the children to do his work so that he can go drinking, it is unacceptable.

Addressing child labour

Pressure to reduce child labour originates mainly from the West and international organisations. The ILO is opposed to child labour (as defined in Table 6.2) and aims progressively to eradicate it (ILO 2002a). To this end it has established an International Programme on the Elimination of Child Labour (IPEC) as well as two international conventions (Box 6.4). Establishing globally acceptable standards is difficult. Convention 182 has been widely ratified partly because Article 3d is sufficiently vague to allow for varying national interpretations (Map 6.1).

Signatories to the ILO conventions pledge to take action against child labour, but, ironically, condemnation of child labour by the World Bank and WTO is likely to have greater impact, as these organisations are able to enforce economic sanctions (Myers 1999). The IFIs are responsible for the structural adjustment programmes that are worsening the condition of the poorest and forcing children into work. Unsurprisingly, their concern for the welfare of working children is treated with considerable scepticism among Third World governments. The debate over 'core labour standards' at the WTO meeting in Seattle in 1999 provoked outrage among Third World delegates who accused the West of seeking to protect its own industries. A Zimbabwean government minister described Western concerns over child labour as a 'ploy by developed countries to protect international markets' (cited in Bourdillon 2000c: 3). Condemnation of child labour is damaging not only to Third World governments, but also threatens multinational corporations that employ children.

In the early 1990s, the media spotlight fell on well-known brands of clothing and footwear, revealing them to be produced using children's labour. The provenance of carpets and footballs available

Box 6.4

ILO measures

IPEC

was launched in 1992 to support governments, employers' and workers' associations and NGOs to develop and implement measures aimed at preventing child labour, withdrawing children from hazardous work and providing alternatives, and improving working conditions as a transitional measure towards the elimination of child labour. The priority target groups are bonded child labourers, children in hazardous working conditions and occupations and vulnerable children: those aged under 12 and girls (ILO 2002b).

The Minimum Age Convention, 1973 (No. 138)

obliges states to set minimum ages for admission to employment which apply across all economic sectors and to all working children, whether they work for wages or on their own account. The general age for entry to employment is set at the completion of compulsory schooling, or not less than 15 years, although children may engage in light work (under 14 hours a week) at 13 and may not undertake hazardous work until they are 18 (or, under strict conditions, 16). Where the economy or educational facilities are poorly developed, governments may set the general minimum age as 14 and for light work 12 years (Martin and Tajgman 2002).

The Worst Forms of Child Labour Convention, 1999 (No. 182)

is intended to prompt governments to take immediate and effective action against forms of child labour that are considered exceptionally harmful to children's welfare, while retaining the ultimate goal of total elimination of child labour. This convention does not distinguish between developed and developing countries and applies to all girls and boys under 18. It does not revise or contradict Convention 138 but singles out as a priority the worst forms of child labour, which it defines in Article 3 as:

a   all forms of slavery or practices similar to slavery, such as the sale and trafficking of children, debt bondage and serfdom and forced or compulsory labour, including forced or compulsory recruitment of children for use in armed conflict;

b   the use, procuring or offering of a child for prostitution, for the production of pornography or for pornographic performances;

c   the use, procuring or offering of a child for illicit activities, in particular for the production and trafficking of drugs as defined in the relevant international treaties;

d   work which, by its nature or the circumstances in which it is carried out, is likely to harm the health, safety or morals of children

(Martin and Tajgman 2002: 122).

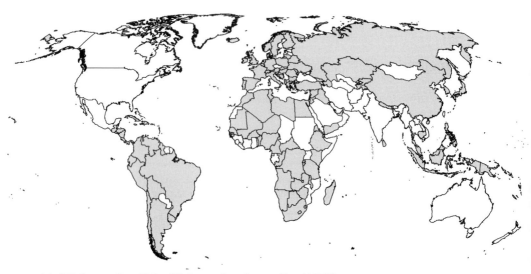

(a) ILO Convention 138 – Minimum Age Convention (1973).

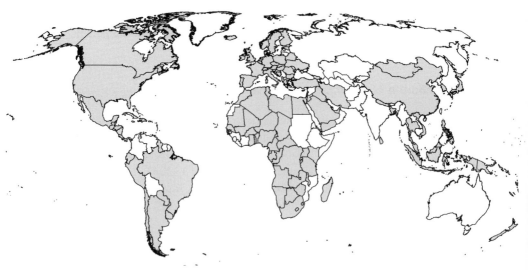

(b) ILO Convention 182 on the Worst Forms of Child Labour (1999).

Map 6.1 *States having ratified ILO Conventions 138 and 182.*

Source: Based on data from UNDP (2003).

Box 6.5

Boycotting the Bangladesh garment industry

Throughout the 1980s and 1990s, garment manufacture in Bangladesh grew rapidly, a large proportion of its output exported to the USA. In 1993, American television broadcast a programme in which Bangladeshi children were seen manufacturing clothes for the US chain Wal-Mart. The resulting consumer pressure persuaded Wal-Mart to cancel its contracts with Bangladeshi factories and prompted US senator Tom Harkin to propose a bill restricting imports to the USA of goods made with child labour. In response, the Bangladesh Garment Manufacturers and Exporters Association (BGMEA) quickly announced the elimination of child labour from the industry (DFID 1999). Children were dismissed, in some cases without pay. While children under 14 were targeted for removal, girls without birth certificates were dismissed on the basis of apparent age, many losing their jobs for want of proof (Seabrook 2001). Although the BGMEA was subsequently involved in negotiations with UNICEF and the ILO to provide education and monthly stipends to the children who lost employment, 75 per cent had been dismissed by the time this became available (Seabrook 2001). Although more adults have since been employed in the industry, ILO/UNICEF research found none of the children excluded from garment manufacture had subsequently attended school and many were engaged in more hazardous and exploitative occupations (Boyden and Myers 1995).

in the West was also traced to child labour. Such revelations led to consumer boycotts. Boycotts have, however, often caused problems (Box 6.5), and have fallen out of favour with NGOs and others concerned about working children's welfare (DFID 1999). Increasingly, international NGOs prefer to promote the use of independently verified company codes of conduct and labelling schemes as ways of addressing child labour.

Inspired by fear of lost custom, many multinational manufacturers have sought to eliminate child labour from the production of goods imported into Western markets. Most global companies favour self-regulation over legislation, and many have adopted codes of conduct. Although usually linked to the ILO conventions, these codes vary in terms of the minimum employment age, whether enforcement is subject to independent monitoring, and what happens when children are found in the company's employment (Kolk and Van Tulder 2002). Few companies give as much attention to providing for the needs of the children they remove as they do to ensuring children are dismissed, leading to suggestions that they are less interested in the children's welfare than retaining market share (Seabrook 2001). Applying codes of conduct to the supply chain is problematic. In garment manufacture, sourcing networks may involve tens of thousands of factories, as well as buying agents, suppliers and subcontractors. If raw material producers are included, monitoring must be extended to agriculture, which commonly employs large numbers of children (Kolk and Van Tulder 2002). Even bringing sub-contracted work into the factory can have side effects. In Pakistan, footballs used to be sewn on a piecework basis by women and children. Under international pressure some employers set up factories with better conditions and remuneration, but the poorest families could not get employment in the factories and lost their source of income (Bourdillon 2000c).

Labelling schemes are used to signal to consumers that goods have been produced to particular standards. These may not guarantee products are made entirely without child labour, but that any children removed from the workforce are provided with replacement income and/or good quality education. *Rugmark*, for instance, was established as a legally binding international trademark in Germany and the USA in the 1990s. Local non-profit foundations in India and Nepal ensure that no illegal child labour is used and the work environment is regulated. Rehabilitation and schooling projects are provided for children displaced from the workforce. The scheme has, however, experienced difficulties in ensuring a comprehensive inspection process (DFID 1999).

Many countries have enacted anti-child labour legislation to fulfil ILO Conventions 138 and 182 and protect overseas markets. Like boycotts, such legislation sometimes displaces children from employment in legitimate organisations to more exploitative occupations. Where children work illegally they cannot complain about working conditions or gain redress if they are injured (Boyden 1991). Reliance on legislation also has the effect of 'penalizing or even criminalizing the ways the poor bring up their children' (Nieuwenhuys 1996: 242), while failing to provide for children to grow up without having to work.

Consumer boycotts, corporate codes of practice, labelling schemes and legislation can effectively address children's work only in the formal sector. In the West, legislation removed children from factories, but it was many years before there was a wider fall in the number of children working. This decline is variously attributed to compulsory education, changes in the perceived roles of children or increases in family income (Nieuwenhuys 1996).

Changing cultural perceptions may result from globalisation, as noted earlier, but are not easy to promote. The eradication of poverty seems even more remote. Many NGOs consider that something should be done to reduce harmful child labour even while people remain poor. Education is seen as both a reason and a means for removing children from the workforce. While compulsory school attendance is difficult to enforce, and causes problems for families dependent on children's work, financial incentives, such as fee remissions, are considered potentially effective. A programme sponsored by the Union of Rural Workers in Brazil, known as *Goats-to-School,* loans goats to families and provides information on goat rearing (Brown 2001). Rather than working in hazardous occupations, children tend the goats. The goats provide milk and meat, and once the loan is repaid by returning a kid for each adult goat, further income from selling young goats pays for schooling.

Far from education conflicting with and thereby contributing to the elimination of children's work, in many cases children need to work in order to attend school. Some schemes view the failure of children to attain an education as a greater problem than their participation in the labour market. One such scheme operating at secondary boarding schools on tea estates in Zimbabwe is *Earn-and-Learn.* Here the company provides schools with relatively good quality facilities (compared with most rural schools), while the government pays the teachers' salaries (Bourdillon 2000a). The students, all aged over 13, are paid to pick tea, receiving the same amount per kilogram as adults. From their pay, the company deducts school fees and meal charges, although these are subsidised. The dormitory accommodation is free. Although the labour is no cheaper than adult labour, it is regular and reliable.

Tea plucking is hard work: the baskets become heavy and the children work seven hours with no break before school. Unsurprisingly, many fall asleep in class. Children are allowed only four weeks' holiday a year to visit their families, and none in the years in which they sit public examinations. Although some students give up and leave, many consider the work no more arduous than that expected of them at home. Most students insist that they join the scheme through their own choice: some because their families could not otherwise afford for them to attend secondary school, but some for other reasons including escaping parental control, the perception that the school offers high-quality education, and the possibility of earning money (Bourdillon 2000a).

In recent years, policy responses to child labour have increasingly recognised that children have opinions, and, in line with the CRC, have the right to be heard. Working children often complain about legislation that is supposedly designed to protect them (Myers 1999). Research with children in Bangladesh, Ethiopia, the Philippines and Central America demonstrates that they are capable commentators on both the hazards of their work and the benefits. Not only do they have no desire for a law that bans them from employment, but most say they would seek ways to contravene such a law (Woodhead 1999). Many children removed from Bangladeshi cigarette factories bribed their way back in, some even scaling the factory wall to re-enter (Seabrook 2001). Those who returned to work illegally were paid less than when legitimately employed. Clearly the children's own perceptions of their interests differed from those of the people who wished to end their exploitation. Where children are seen as participants in the fight against exploitation, not opponents (Miljeteig 1999), there is a stronger possibility that interventions will operate in their interests. For the first time in its history, the ILO involved working children in consultations about the 1998 convention. The children expressed themselves in favour of 'work with dignity and appropriate hours', and time for education and leisure, but opposed to boycotts of products made by children (White 1999: 140). The trade union movement has traditionally opposed child labour, as, in the past, it opposed the labour of women. One day, trade unions might recognise that children are legitimate employees, who not only have interests in the workplace that need representing, but who are able to participate in securing benefits for workers of all ages.

Youth employment and unemployment

While in some quarters the employment of young people is condemned, in others the lack of employment among (in many cases the same) young people arouses concern. Particularly for male youth, leaving school without immediately entering training or work is commonly considered unacceptable. Where secondary education is not widely available, school leavers may be quite young. Youth unemployment is often seen in terms of moral crisis, the blame cast on young people themselves, who are regarded as idle or lacking the right attitudes, while in practice jobs may be extremely hard to obtain.

More than 40 per cent of the world's unemployed are youth (United Nations 2002). In most parts of the world youth unemployment exceeds that among older people (Figure 6.2). The unemployment rate provides only a partial picture of young people's employment difficulties. 'Increasingly, the distinction between employment and unemployment has lost much of its meaning, as young people move in and out of informal activities where neither term has any real relevance' (United Nations 2002: 4). Many young people are underemployed, meaning they have insufficient work to keep them fully occupied, or provide an adequate income, or they have poor-quality informal sector employment. Others are classed as outside the workforce, because they have become disillusioned with unsuccessfully seeking work (O'Higgins 2001).

There are several reasons why youth unemployment is high. Some blame population growth, but in practice there are many more significant factors (O'Higgins 2001). Youth unemployment rates are closely correlated with both adult unemployment and economic growth, but the youth labour market (like child labour) is fairly

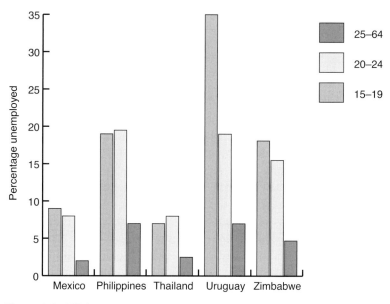

Figure 6.2 *Official unemployment rates by age group.*
Source: Adapted from O'Higgins (2001).

distinct from the adult labour market. Young people cannot compete for jobs that require high-level skills or experience, but perceived attributes of young people, such as adaptability, are valued in some employment situations (O'Higgins 2001).

Youth unemployment is particularly sensitive to economic decline. Structural adjustment led to falling public and private sector employment in many countries in the 1980s and 1990s, affecting young people in two ways. First, many employers stopped recruiting, adversely affecting young entrants to the job market. Second, young people are easy targets for retrenchment. They have generally received less investment in training than older employees; redundancy payments cost little; and they are less likely to be covered by employment protection legislation (O'Higgins 2001).

The labour market in structurally adjusting Central Asia offers very uncertain prospects to school leavers (Falkingham 2000). In Soviet-era Kazakhstan the opportunities available to children and youth reflected the social standing of their parents, with the best universities effectively reserving places for those whose parents held high office (Rigi 2003). Working-class youth, nonetheless, had access to all levels of education, secure jobs and sports and cultural facilities, although acquiring an apartment or a car took many years. In the post-Soviet era, Kazakhstan has experienced deepening social polarisation. Although the children of the elite can secure university places, good jobs and indulge in conspicuous consumption, most young people can find only menial informal sector employment. Yet, like their wealthier peers, these young people aspire to consumerism, many resorting to illicit activities to access goods (Rigi 2003).

Unemployment is concentrated among certain groups of young people, particularly the poor. There is no clear global pattern regarding the gender distribution of youth unemployment, but labour force participation among women is usually lower for all age groups. In urban Tanzania in 1990–1991, unemployment among 20–24-year-old women, at 25.7 per cent, was almost double that among young men (a much bigger difference than for 15–19-year-olds). Whereas most of the young men who were unemployed in their late teens subsequently found employment, young women faced longer-term difficulties. Women who succeeded in entering employment often had less desirable jobs, and underemployment among women was more than double that for men. In many Third World countries, young men outside the labour force are pursuing education; young women are more likely engaged in housework, a situation that exacerbates young women's labour market disadvantage. However, even in Jamaica, where educational attainment is higher among women than men, women are more likely to be unemployed (O'Higgins 2001).

In rural areas of the Philippines, female youth unemployment is high, as women are less likely than men to find agricultural work (and female agricultural work often remains unrecorded in employment statistics). However, significant numbers of young underemployed Filipina women are finding work overseas, mainly as domestic helpers (Santa Maria 2002). 'In the Philippines, it is daughters who can be expected to meet the family needs more than sons, hence, it is they, rather than sons, whom the family entrusts with the migration task' (Lauby and Stark 1988: 486). Elsewhere, young women's labour force participation is also increasing. In Egypt, economic hardship has drawn many into civil-service jobs. These are stable, require short hours and are therefore regarded by parents as respectable, but only for a temporary period preceding marriage (Booth 2002).

It is often claimed that in Third World countries youth unemployment is concentrated among those with higher levels of education. While some data appear to support this claim, there are numerous counter-examples (O'Higgins 2001). In Namibia, for instance, unemployment among university graduates in 1997 was 3.7 per cent, compared with 39 per cent among those with only primary education. Even in Malaysia, where officially unemployment for the least educated young men is 9.4 per cent and for the most educated 14.8 per cent, when 'discouraged workers' are added to the unemployed the rates change to 29.8 per cent for the least educated and 21.7 per cent for the most educated. The data are also skewed where unemployment by educational level is cited for a population as a whole, as the higher educational levels tend to be concentrated among younger people, for whom unemployment levels are higher regardless of education. Furthermore, although data are often quoted that reveal that it takes longer for a more educated young person to find employment, this may reflect the luxury of the elite to wait for a high quality job (O'Higgins 2001).

Educated under- and unemployment is common in India, particularly since structural adjustment reduced the availability of white collar and professional jobs. In 2000, only about one in 24 job seekers in India obtained formal sector work (Jeffrey *et al.* 2003). In Uttar Pradesh, where the World Bank insisted on a two per cent annual cut in public sector employment, young men from the *Dalit* ('untouchable') castes find job-seeking particularly difficult, being discriminated against and lacking useful social contacts or money for bribes (Jeffrey *et al.* 2003). As educated men they are unwilling to work as labourers, even though they could earn more money this way. Education has given them self-respect, and, while they resent the wider options available to the uneducated, labouring would be demeaning. Most describe their current informal employment positions as provisional (Jeffrey *et al.* 2003).

Young people from the Philippines also prefer to work outside agriculture. As a consequence, a shortage of labour in the agricultural sector coexists with relatively high levels of youth unemployment. In some villages, as many as 20 per cent of adults, mostly aged 15 to 30, class their occupational status as 'standby' – they have no employment, no direct source of livelihood and are not in education but are waiting for a suitable opportunity. Unlike in the past, economic circumstances allow them to avoid agricultural work. Work in factories, offices or shops, in the cities or overseas is seen as preferable, bringing greater economic and cultural capital. Education contributes to this cultural reworking. Local secondary school curricula are geared to the needs of the export sector, with short technical courses on electronics, dressmaking and computing. At one large high school, of the 40 per cent of graduates who enter the workforce, three quarters find work at the Cavite Export Processing Zone. A survey among students revealed that none wished to become farmers and farming was seen very negatively by over three-quarters. Economic change has brought new opportunities that enable young people to be selective in their choice of employment (Kelly 1999).

Youth employment policy

> Youth unemployment can lead to marginalization, exclusion, frustration, low self-esteem and, sometimes, to acts that create burdens on society.
>
> (United Nations 2002: 4)

The UN, in relation to youth, has focused centrally on school to work transitions, and the ILO has, since 1935, sought to reduce youth unemployment. This partly reflects the familiar concern with young people 'as and at risk'. Many young people remain unemployed for long periods, which can jeopardise their later job prospects, as well as encouraging crime and drug use. High youth

Box 6.6

Tackling youth unemployment in Jamaica

Jamaica has high levels of youth unemployment, related in part to structural adjustment in the 1990s. Economic growth averaged 5 per cent a year in 1986–1990, but following deregulation of the exchange market in 1991 fell to an average 1 per cent a year through the 1990s. The slowdown affected the young disproportionately, such that by 1995 the ratio of teenage unemployment to that among people over 35 was 14.5:1 and for young adults 9.3:1 (up from 4:1 and 3:1 respectively in 1980).

The Human Employment and Resource Training (HEART) Trust was launched in 1982. Residential and non-residential academies were established providing off-the-job vocational (70 per cent) and academic (30 per cent) training for young people who had completed Grade 9 of secondary education. Of seven academies, six concentrate on a specific skill (hospitality, cosmetology, commercial business training, construction, garment manufacturing and agriculture) and the other is a multi-skills centre. Most courses last a year and students, who may be resident or non-resident, receive a small allowance. In 1995–1996, nearly 4,000 students were enrolled. The academies' graduates have significantly higher rates of employment than their peers and also earn higher wages, although the garment manufacture and agriculture academies have been less successful. In addition to the academies, HEART runs a schoolleavers' programme providing job placements with on-the-job training for those leaving school after Grade 11. About 3,000 young people undertake one- to three-year courses. This too increases the likelihood of obtaining employment. HEART is funded from mandatory employer contributions, for which firms receive a credit for each trainee accepted, thereby encouraging them to participate, along with contributions from international donors.

Although HEART has had some success, it has concentrated on young people who already have secondary education, rather than those for whom access to employment is most difficult. There is also something of a mismatch between the skills provided and those demanded by the few growing sectors of the Jamaican economy. Currently, demand for skilled and educated labour in Jamaica is low, and until more jobs are created that match the aspirations of educated young people the youth employment problem will remain. Furthermore, the scale of youth unemployment in Jamaica is such that HEART is too small to have a significant impact.

Source: O'Higgins (2001).

unemployment may also lead to alienation and social unrest (O'Higgins 2001).

The UN, World Bank and ILO have launched a Youth Employment Network to address youth unemployment, highlighting four areas that need attention: employability (education and training); equal opportunities (for young women); entrepreneurship (facilitating creation of small enterprises) and employment creation (focus on macroeconomic policy) (United Nations 2002). A broad distinction exists between interventions aimed at promoting waged employment by offering vocational training and sometimes subsidised work placements and those that encourage young people to become self-employed by offering training in business methods, access to credit or grants, and work space. Training may be provided 'on the job' or between school and fulltime work, and may take place in the workplace, in vocational training institutions or a combination. The use of institutions does not preclude employers' involvement in training (see Box 6.6).

In practice, most employment interventions neglect the most disadvantaged young people, helping only those best equipped to help themselves (O'Higgins 2001). Programmes seldom fit the needs of rural youth, those on low incomes, the youngest, least educated and often women (Fawcett 2001). The lack of success of post-Apartheid South Africa's youth programmes is attributed partly to their focus on high-skill formal employment (Chisholm *et al.* 1997). It is increasingly advocated that training be provided in skills relevant to both formal and informal sectors (Fawcett 2001). Much training for informal sector work takes place within the informal sector, although the most productive businesses in Kenya's informal (*jua kali*) sector are those that benefit from training and experience in the formal sector (King 1996).

Conclusions

The involvement of children and youth in work is subject to moral discourses that are culturally variable and often inconsistent. The age at which children begin work in different societies and different families is affected by a range of factors, both economic and attitudinal. In some parts of the world it is no more scandalous to value a child for (among other things) his/her economic contribution to the household than to value a parent in this way: children are valued for what they do rather than for simply existing. This gives them a real role, and sense of worth, as actors, not just possessions. While some children benefit from the work they do, others are doubtless engaged in occupations that jeopardise their welfare, either immediately or in their future lives. Economic systems seldom work in the interests of the poor and those who are employed are often in some sense 'exploited'. However, despite campaigns by NGOs and international organisations, it is increasingly recognised that depriving children of the opportunity to work can be damaging. There is a need to strengthen the bargaining power of those who sell their labour, either to ensure that they are better rewarded and work in better conditions, or that they are able to secure their livelihoods in alternative ways. Yet, while some child-oriented organisations recognise that children are economic actors, this recognition seldom extends to other organisations. For instance, very few small business support or microfinance organisations work directly with children, although assistance to family businesses can impact significantly on children's workloads (Moore 2000).

For those leaving school in their teens, in contrast, there are numerous programmes aimed at fitting them to the labour market. This reflects widespread concerns about youth unemployment. Although, for structural reasons unemployment among youth is common globally, moral discourse tends to blame the individual. Public action to address youth unemployment is also related to fear that young people will become disillusioned or dissatisfied, and consequently pose a threat to society.

Key ideas

- Much international thinking concerning young people and work is founded on the assumption that children should remain in fulltime education until at least their late teens and then make a transition to fulltime work.
- Across the world very many children work: in most societies children have always worked, and in many cases their work is necessary to household survival.
- The work children do varies greatly in terms of type of activity, labour relations, setting, remuneration, degree of hazard and time commitment.
- Some children engage in work that is hazardous to their health or interferes with their education, while for others work is a more positive experience.

- Interventions aimed at 'rescuing' children from work sometimes result in them working in more hazardous illicit occupations: supporting working children's rights may be more helpful than depriving children of the right to work.
- Youth unemployment is considered a growing problem in many countries, particularly where growing numbers of children are leaving secondary school.
- Programmes are designed to facilitate entry of young people into employment or self-employment.

Discussion questions

- Should children have the same rights as adults to earn money to support their own livelihoods?

- Is it appropriate to establish international guidelines on the minimum age that children might engage in particular occupations? How might this be done?

- Do youth training schemes serve the interests of young people themselves, or of a wider society that fears young people?

Further resources

Books

Bourdillon, M. (ed.) (2000) *Earning a life: working children in Zimbabwe*, Harare: Weaver Press – a collection of research-based essays about children working in various occupations.

Nieuwenhuys, O. (1994) *Children's lifeworlds: gender, welfare and labour in the developing world*, London: Routledge – examines children's work, mainly within a family context, in rural Kerala, India.

O'Higgins, N. (2001) *Youth unemployment and employment policy*, Geneva: International Labour Office – investigates the causes of youth unemployment and effectiveness of different policies around the world.

Seabrook, J. (2001) *Children of other worlds: exploitation in the global market*, London: Pluto – relates child labour to poverty, both in nineteenth-century Britain and in the Third World today.

Journals

There are no journals that focus exclusively on working children and youth, but *Childhood* sometimes publishes relevant research articles. See the 1996 special issue on street and working children.

Websites

Anti-Slavery International http://www.antislavery.org/ campaigns against child labour.

Global March Against Child Labour http://globalmarch.org/index.php is a campaigning NGO.

The International Congress of Free Trade Unions http://www.icftu.org/focus.asp?Issue=childlabour&Language=EN puts its case against child labour.

ILO's IPEC pages http://www.ilo.org/public/english/standards/ipec/index.htm provide information including the texts of ILO Conventions 138 and 182.

Save the Children http://193.129.255.93/labour/index.html has a range of downloadable publications on child labour.

UNICEF http://www.unicef.org/protection/index_childlabour.html has a range of information and publications on child labour.

7 Children in especially difficult circumstances

Key themes

- Children in armed conflict
- Street children
- Sexual exploitation of children
- Children affected by AIDS
- Children with disabilities
- Children in institutions.

Many children and youth in the Third World live in circumstances that would generally be considered 'difficult' in the West. Numerous children suffer periodic or chronic ill health; many do not attend school; and a sizeable proportion are working at an age when children in the West enjoy plentiful leisure time. A minority of children, however, experience circumstances that are more extreme, and more difficult (Table 7.1). In the 1980s, UNICEF began to focus on the needs of children in situations of warfare; those with disabilities; children exploited for their labour and for commercial sexual gratification (UNICEF 1996). For these it coined the term 'children in especially difficult circumstances' (CEDC). The purpose was to provide children in pre-defined circumstances assistance appropriate to their situations, largely in relation to assuring their survival, health and education (Majekodunmi 1999).

Children in difficult situations such as these have dominated the media spotlight and attracted both academic and NGO attention.

Table 7.1 *Approximate numbers of children in difficult circumstances*

	Approximate number
Workers	250,000,000
Malnourished (under five)	174,000,000
Out of school (primary school age)	140,000,000
Displaced by conflict	30,000,000
Orphaned by AIDS	13,000,000
Dying of preventable disease (each year)	12,000,000
Living on the streets	8,000,000
Living in institutions	7,000,000
Sexually exploited	1,000,000
Serving as child soldiers	250,000

Sources: Based on Lansdown (2000); USAID/UNICEF/UNAIDS (2002); Tolfree (1995).

While far from typical of Third World children, CEDC merit attention, in part because of the high profile they have in the Western imagination of the Third World and in NGO policy. It is important that such children's lives are understood and they are not merely on the receiving end of emotional but uninformed responses.

Since the mid-1990s, UNICEF has employed the term 'child protection' as an umbrella term for its actions towards CEDC. Specific concerns have shifted over time. At one time, street children were dominant, being highly visible, but not necessarily the most

---

Box 7.1

UNICEF circumstances for priority action in child protection

- Children in forced and bonded labour;
- Children without primary caregivers (ranging from those separated from families by war, orphaned by AIDS or detained in state institutions);
- Children who are trafficked;
- Children who are sexually exploited;
- Children who are used as soldiers;
- Children subjected to violence outside armed conflict (including female genital cutting).

Source: UNICEF (2003a).

needy (Connolly and Ennew 1996). The current UNICEF priority areas are listed in Box 7.1. Many of these circumstances are addressed in Article 39 of the CRC:

> States Parties shall take all appropriate measures to promote physical and psychological recovery and social reintegration of a child victim of: any form of neglect, exploitation, or abuse; torture or any other form of cruel, inhuman or degrading treatment or punishment; or armed conflicts.
>
> (United Nations 1989)

Children may experience difficult circumstances for a number of reasons, but in most cases the root problem is an underlying context of acute vulnerability. In many cases, it is the livelihood strategies that particularly impoverished and vulnerable children and families employ that place children in especially difficult circumstances. These difficult circumstances are also often interrelated. As a result of war, for instance, children may be sexually exploited, separated from their parents, live on the streets, or be incarcerated for crimes committed as a child soldier. Those living on the streets are more vulnerable to being trafficked or recruited as soldiers. This chapter focuses on six groups of young people in especially difficult circumstances, four of whom have had considerable media exposure in the West (those affected by war, including child soldiers; those working in the commercial sex industry; those living on the streets; and those affected by AIDS) and one group who have received much less attention (young people with disabilities) along with children living in the institutions provided to care for CEDC that often represent difficult circumstances in themselves.

Children in armed conflict

Young people are affected by armed conflict in ways that are different, and sometimes worse, than adults. Historically, most societies have sought to protect children in war. The Acholi people in Uganda, for instance, avoided attacking children in order to facilitate post-conflict reconciliation (Otunnu 2000). Under international law (Geneva Protocols, 1977), children under 15 are entitled to special treatment: they should not be arbitrarily killed, tortured or ill-treated; should have priority in receiving food and shelter and be kept with their families where possible (Kuper 2000). Such principles are not universally adhered to. In some cases, children are deliberately targeted: their iconic value as representatives of both innocence and society's future renders them potent in pressurising populations. Children have been employed as human shields, or killed by terrorists. As many as 90 per cent of war victims today are civilians, around half of them children (Majekodunmi 1999).

Children are affected in many ways by conflict, and not all share similar experiences. Pictures of bewildered infants with limbs shattered by landmines are far removed from the lives of most young people in war. In some conflicts, children are direct victims. Most are not deliberately targeted, but are vulnerable to injury. Children may, for instance, fail to spot landmines because they are short, or do not recognise what they are, or may be enticed by curiosity to investigate. Mines designed to maim adults kill 10,000 children a year, often long after the conflict is ended (Galperin 2002).

Most children who die in war are killed indirectly, by, for instance, the destruction of hospitals and water purification plants and disruption of agriculture and food distribution (Carlton-Ford *et al.* 2000). Seven of the 10 countries with the highest child mortality rates have recently experienced serious armed conflict (Galperin 2002). Most excess deaths are in the one to four age group, the most vulnerable to malnutrition and disease, but death rates also rise among five- to nine-year-olds during war (Galperin 2002).

Relatively few children are killed or injured by conflict, but many are secondary victims or observers, experiencing, for instance, damage to family property or death or injury to relatives (Boyden and Gibbs 1997). About a million children have been orphaned in conflicts in the past decade, and about 12 million children are displaced from their homes (Otunnu 2000). Many live in camps for internally displaced persons (IDP) or refugees, which are usually overcrowded and unhygienic, exposing children to disease.

An estimated 3.5 per cent of displaced children worldwide are unaccompanied, having been separated from parents or other adults who would normally care for them (Machel 2001). These children have the least access to education, health care and nutrition. Among the estimated 24,500 unaccompanied refugee children in Burundi in 1996 were some who, having growing up amid people of varied national and linguistic backgrounds, employed mixed vocabularies making it impossible even to identify their country of origin (Majekodunmi 1999).

Rape is relatively commonplace during war, in refugee camps and elsewhere (Majekodunmi 1999). UN peacekeepers and other military personnel have contributed to the rise in child prostitution in Cambodia, Liberia and Angola (Machel 2001). The heightened HIV prevalence among soldiers puts children at risk of AIDS (Galperin 2002).

Conflict also disrupts education. Although the International Criminal Court classifies targeting schools and hospitals as a war crime (Otunnu 2000), schools are sometimes deliberately targeted; others are destroyed inadvertently or closed down. Even where schools remain open, parents may fear to send their children. During prolonged conflicts an entire generation may miss out on education, denying them subsequent administrative or political roles (Majekodunmi 1999).

Considerable attention has recently been paid to war's psychosocial impacts. Children exposed to violence may experience behavioural problems (Table 7.2). Children are not necessarily, however, more vulnerable psychosocially than adults. Young children may become anxious, particularly if their parents become overprotective, emotional or unusually authoritarian, or where social networks are disrupted (Boyden and Gibbs 1997), but the view of children as especially vulnerable is rooted in a Western notion that children need protection from real life. Only a small minority of children suffer long-term mental health problems when exposed to conflict situations, and most find ways of coping with stress (Boyden and Gibbs 1997; Loughry and Flouri 2001). Although war situations are often far removed from the safe, stable family environment advocated for children (Boyden 2003), where conflicts are prolonged, war is experienced as mundane (Boyden 2000). Furthermore, growing up amid change and contradiction can be a source of resilience rather than risk (Boyden 2000).

Children differ in their resilience (Box 7.2) and in many cases the differences between individual children are wider than between children and adults (Boyden and Gibbs 1997). Children employ multiple coping strategies, some of which are unavailable to adults such as scavenging in areas controlled by security forces (Boyden 2000). Many take on responsibilities as carers and income earners earlier than they otherwise would (Boyden 2000). In such situations, it is better for children to learn to protect themselves than always to be protected by adults (Boyden 2000).

Table 7.2 *Common responses of children to disasters*

Toddlers	School-age children	Adolescents
Regression in behaviour	Marked reactions of fear and anxiety	Decreased interest in social activities, peers, hobbies, school
Decreased appetite	Increased hostility with siblings	
Nightmares		Inability to feel pleasure
Muteness	Somatic complaints (e.g. stomach aches)	Decline in responsible behaviours
Clinging	Sleep disorders	Rebellion, behaviour problems
Irritability	Decreased interest in peers, hobbies	Somatic complaints
Exaggerated startle response	Social withdrawal	Eating disorders
	Apathy	Change in physical activities (both increase or decrease)
	Re-enactment via play	
	Post-traumatic stress disorder	Confusion
		Lack of concentration
		Risk-taking behaviours
		Post-traumatic stress disorder

Source: Jabry (2002).

Box 7.2

Factors affecting children's resilience

- Age: infants are physically and emotionally dependent; older children's dependence is socially determined and variable.
- Gender: young boys may be more vulnerable than girls, whereas older boys are more resilient.
- Past experience: resilience may be learned through rites of passage or previous responsibilities.
- Temperament and cognitive capacity: particularly the ability to think critically.
- Social networks and economic resources.

Sources: Boyden and Gibbs (1997); Boyden (2000).

Various measures have been employed to mitigate war's impacts on children. Humanitarian ceasefires permit inoculation and sometimes evacuation of children from conflict zones (Otunnu 2000). Beyond securing basic needs, NGOs seek to address the separation of families, sexual exploitation and violence, forced labour, corruption and banditry (Boyden and Gibbs 1997). Education and recreational activities in IDP camps encourage a sense of normalcy (Jabry 2002). Vocational training provided to young Burundian refugees occupies their time and provides income-generating skills of use both in the refugee camp, and, more importantly, when they return home (Lyby 2001). In responding to the psychosocial impacts of conflict, Western psychotherapeutic models are increasingly seen as culturally inappropriate and are giving way to approaches focusing on social reconstruction, reconciliation and healing, attempting to rebuild social networks, reinstate community structures and develop mechanisms for justice and retribution (Boyden 2003).

Many humanitarian responses to war treat children as passive victims. Understanding children to be resourceful allows interventions to build on their strengths rather than assuming they are dependent on adults and outside expertise, which might not be available (Boyden 2003). Children can, for instance, care for younger children, including unaccompanied minors (organising entertainment as well as more mundane tasks), gather wood, distribute water, work in gardens, cook, help in clinics and make clothes (Jabry 2002). Bringing together children from opposing sides in a conflict to discuss peace, may even improve the prospects for future resolution (Jabry 2002).

Child soldiers

Children are involved in conflicts not only as victims but also as soldiers. Soldiers are usually young, but over the past decade the very young age of many recruits has aroused international alarm. Between 1994 and 1998, 35 armed conflicts involved soldiers under 15 (Hoiskar 2001). Interviews with child soldiers across East Asia and the Pacific found most were recruited aged between 12 and 14, and one was only nine (UNICEF 2002b).

The image of the child soldier is disconcerting, disrupting popular notions of children as innocent and vulnerable. The international community has tended to define child soldiers as victims rather than perpetrators, absolving them of responsibility for their actions.

International law forbids recruitment of soldiers under 15, and since 2002 an Optional Protocol to the CRC on the Involvement of Children in Armed Conflict prohibits the involvement of anyone under 18 in hostilities.

International law notwithstanding, children are recruited by both government and rebel forces for various reasons. In many Third World countries, children are plentiful; the proliferation of small weapons, including light semi-automatic rifles such as AK47s, has overcome children's physical disadvantages; and children are perceived to be energetic and undemanding (Rabwoni 2002). Some children are recruited forcibly (for instance, by kidnapping), others are handed over by families or communities, often under intimidation, but most volunteer (Jok 2003). Reasons for volunteering include intimidation and fear (Hoiskar 2001); the chance to escape extreme poverty or unemployment (de Waal 2002); quest for adventure inspired by films and books (Rabwoni 2002); or a political cause. NGOs emphasise the role of poverty and force, viewing children as incapable of forming political views: '[u]nlike adults . . . children cannot be expected to distinguish between competing causes' (UNICEF 2002b: 23). Yet, while very young soldiers may lack sufficient understanding of the issues to have informed opinions, and young people can be easy prey to those promising radical reform, '[t]o treat them merely as deluded or wayward children would be to depoliticize their project and to fail to address their grievances' (de Waal 2002: 22). Palestinian suicide bombers are predominantly young, but accusations that they are manipulated with the promise of martyrdom meet with denial by young people who express a commitment to political Islam (Khashan 2003). Young people's influence in rebel movements varies: while Sierra Leone's Revolutionary United Front was arguably youth-led, Liberia's young rebels were in reality manipulated by middle-aged powerbrokers (Ellis 2003).

Young people's experiences of military life vary considerably. Some never experience front-line fighting and are employed only as cooks, cleaners, guards, spies, lookouts and messengers. Girls are often sexually exploited (Hoiskar 2001). Many girls and boys are quickly introduced to armed combat. Because they are numerous and have received little training, children are often considered expendable and may be assigned the most hazardous tasks (Hoiskar 2001).

Many child soldiers are required to kill. Some act with less restraint than adults, for instance shooting prisoners of war in revenge for the deaths of their families (Rabwoni 2002). Although some are traumatised by the experience, participation in killing is often experienced as an initiation into adulthood (Ellis 2003), and by no means all child soldiers wish to be demobilised. For girls from rural villages, army life can be personally emancipating. Research by UNICEF (2002b) found that many child soldiers thought recruitment of those under 18 should be prohibited, but some believed any child old enough to wield a weapon should have the right to be a soldier.

The militarisation of young people has impacts that extend beyond child soldiers themselves. South Sudan has been engulfed in civil war since 1983. The current generation of military recruits has grown up in conflict, and exhibits an 'ultra-masculinity', which emphasises aggression, competitiveness and the censure of emotional expression, along with a sense of entitlement to the domestic and sexual services of women (Jok 2003). When a conflict ends, such problems do not disappear. Demobilised soldiers, particularly those recruited young, may find it difficult to reintegrate into society.
If the post-conflict government fails to deliver what they believed they were fighting for, or older politicians hijack their place, they may quickly become disillusioned (Rabwoni 2002).

Demobilisation of child soldiers is often problematic. In many parts of the world, young soldiers evoke little sympathy. In parts of West Africa, youth are feared and believed to be out of control, fuelled by frustration, lack of education and easy access to both drugs and weapons. While children may have killed, tortured and sexually abused civilians, including other children, international law requires that they be treated in the same way as child civilians (Kuper 2000). Although some are incarcerated, most child soldiers are put through rehabilitation programmes (ILO 2003) (Box 7.3), which may not alleviate the fears of the communities to which they return.

Defining all military personnel below 18 as 'child soldiers' unhelpfully elides nine-year-old combatants with those aged seventeen, while unrealistically distinguishing 17-year-old 'victims' from 18-year-old 'perpetrators'. Focusing on child soldiers as 'innocent victims' fails to illuminate the involvement of young people in their mid-teens and also the continuity of motivation and experience across the 18 barrier. There is also a danger in conceptually separating child soldiers from their broader contexts of political crises (Rabwoni 2002).

Box 7.3

Demobilising child soldiers in northern Uganda

Northern Uganda has known conflict since 1986. From 1994, the Sudanese government began supplying large volumes of arms to the Lord's Resistance Army (LRA), creating a demand for soldiers to wield them. While some young Acholi people volunteered, around 80 per cent of the new recruits were abducted, most aged under 18 and almost half aged 11 to 16. By 1999, it was estimated that approximately 10,000 girls and boys had been abducted for use as soldiers. The recruits were taken across the border to camps in Sudan. With as many as 50 per cent escaping and many killed in combat, there is a constant need to replenish the ranks.

The Ugandan government is faced with a dilemma in determining how to treat those who return from the LRA having committed atrocities. While there are fears that to allow people to go unpunished for serious crimes might encourage further crimes in future, there is also recognition that exacting vengeance tends to perpetuate a spiral of violence. A potential solution has been found in appealing to the CRC and the African Charter on the Rights and Welfare of the Child, both of which are incorporated in the Ugandan Children's Statute. As most soldiers were recruited when under 18, there is agreement that it is inappropriate to treat them as adult criminals. However, 15 to 20 per cent of returnees are over 18, and perhaps 6 per cent were over 18 when the conflict began. Among the Acholi, however, a child is defined not by chronological age, but by a range of social factors. The authorities in northern Uganda and the Acholi people have been treating all returnees as children and handing them over to NGOs for family tracing, psychosocial counselling and reintegration programmes. While this is non-confrontational, it poses problems for NGOs whose programmes are geared to the needs of those under 18. Older soldiers sometimes seek to reassert their authority over the younger children, and resent being expected to engage in activities designed for children. The establishment of reintegration programmes designed for adults has not, however, found favour with the authorities who rely on the external support that can be mobilised around the notion of child soldiers.

Source: Mawson (2000).

Street children

For centuries, children have lived and worked on city streets, but as recently as 1979 no NGO programmes existed for street children (Rosemberg and Andrade 1999). In the early 1980s, reports that 100 million children lived on the streets worldwide alerted international NGOs. Empirical research uncovered a less alarming picture.

Reports of 100,000 street children in Mexico City were confounded by a survey that found only 11,172, of whom over 10,000 were living at home (Black 1993).

Street children represent highly visible challenges to idealised (Western) notions of childhood. The term 'street child' is a recent invention, replacing local variants. Juxtaposing 'street' and 'child' to generate a label for an object of concern implies children's presence in public space is illegitimate. As 'poor children in the wrong place' (Scheper-Hughes and Hoffman 1998: 358), street children make the public aware that poverty exists and affects children. Also conflicting with the Western childhood ideal is the apparent absence of adult supervision. Children are assumed to be abandoned, and blame is cast on their families. Parents are represented as having too many children, indulging in alcohol and drugs, raising children without fathers and having multiple casual sexual partners (de Moura 2002). Studies in Maputo, Mozambique and Asuncion, Paraguay, revealed street children were not drawn disproportionately from one-parent families (Boyden 1991). A more important factor is economic marginality (Rizzini and Lusk 1995), but this receives much less attention. The role of children's own decision-making is also seldom acknowledged (Young 2004b).

While street children's families are represented as negligent, children themselves are portrayed as both victim and deviant. Discussions of street children tend to focus on deficits – lack of shelter, food, education and health care (de Moura 2002). Street children are represented as alien to mainstream society, defying moral values by indulging in drug use, sexual promiscuity, prostitution and crime. They are seen as feral and untamed; not fully responsible for their own behaviour (de Moura 2002); inevitably psychologically damaged and destined for failure as adults (Ennew 1994).

These representations of street children both overdraw the distinctiveness of street children and homogenise a wide range of experience. Street children are a varied population, whose experiences differ according to, among other attributes, age, gender, ethnicity and class (Box 7.4). UNICEF differentiates between street children through a three-fold categorisation: 'candidates for the street' (poor children who spend time hanging out or working on the streets), 'children on the street' (those who work on the street, but usually sleep at home) and 'children of the street' (those who live on the street without family support) (de Moura 2002). Although widely

Box 7.4

Gendered street-child cultures in Yogyakarta, Indonesia

In 2000, an estimated 1,600 children were living on the streets of Yogyakarta, an impoverished city in south-central Java, of whom 500 were girls aged 4 to 16. Their presence was partly attributable to Indonesia's pursuit of neo-liberal economic development based on foreign investment and industrialisation, which brought millions into the cities but witnessed a growing gap between rich and poor. As the poor became increasingly marginalised, children drifted to the streets, a process that accelerated following the 1997 financial crisis.

Not all children explain their presence on the streets in terms of poverty. Some claim to have left home because they felt unloved and were beaten, their fathers drank too much, they felt under excessive pressure to do well at school, their parents were absent or had separated, or they wished to avoid hostile stepparents. Others blame the influence of friends or attraction of street children's subcultures. Some girls sought to escape the 'stifling' strictures of the traditionalist gender ideology known as *state ibuism*: devotion to domestic chores, always being home by 9.30p.m., and prohibitions on smoking, alcohol and pre-marital sex.

Street children are widely shunned. Traditional Javanese society condemns their loss of kinship ties. The Western-influenced middle classes resent being confronted with dirty homeless children in the city centre. Under neo-liberalism public space is increasingly privatised, excluding children from, for example, shopping malls. Street children also disrupt the image of progressive modernity the state wants to portray to foreign investors. In periodic 'cleansing operations' children without identity cards are removed from the streets. They are detained, subjected to verbal and physical abuse, their means of livelihood (guitars for busking, goods for sale) confiscated, and some have been shot attempting to flee from the police.

Girls are less visible on the streets than boys, and although some go to the streets to avoid gender discrimination, once on the streets they are doubly excluded. Despite their own alienation from mainstream society, street boys subscribe to conventional patriarchal attitudes. They abuse girls, refuse to acknowledge them as street children, but liken them to prostitutes. Public attitudes preclude girls from employing boys' livelihood strategies: shining shoes, scavenging for goods to recycle, busking, selling handicrafts, 'parking' cars and begging. Girls are therefore dependent on boyfriends, receiving material support in exchange for sex, although few work as prostitutes.

Street children resist their exclusion through subcultures. Boys claim areas of the city as their own, with groups naming themselves after the places where they work, sleep and hang out such as the main street, train station, bus station, market and public toilets, and using the *Tikyan* private language. Street girls resist their marginalisation too, carving out their own niche spaces, as they are excluded by many of those dominated by boys. Girls also drink, smoke, take drugs, have tattoos, scarify their arms with razor blades, wear boys' clothes and talk and act tough, refusing conventional notions of femininity.

Sources: Beazley (2000b, 2002).

used, this categorisation is problematic. First, it suggests a causal progression, but working on the streets does not inevitably lead to sleeping on the streets (Rosemberg and Andrade 1999). Second, children move fluidly between categories (Godoy 1999). In Rio de Janeiro, for instance, many children from the peripheral slums work on the streets, and avoid a long and expensive daily commute by sometimes staying in vacant buildings, shelters, doorways or pavements (Rizzini and Lusk 1995). Furthermore, most Latin American street children retain links with their families, and some live on the street *with* their families (Rizzini and Lusk 1995).

Children living on the streets exercise more choices than children in other environments (Connolly and Ennew 1996), and survive by employing a range of coping strategies. Indeed, they are attracted to city centres partly by income-generating possibilities: selling newspapers, shining shoes, guarding cars, cleaning windscreens or guiding tourists. They know the urban environment well and carve out niches in which to perform daily activities: work, leisure, washing, eating and sleeping (Young 2004a). Many experience health problems, particularly skin diseases, respiratory infections, intestinal parasites and injuries associated with their lifestyles (Ennew 1994). Many take drugs or inhale glue to stave off hunger, escape anxieties and sleep. Unprotected sex, for survival or pleasure, also carries risks for children whose priority concerns are food, money and clothes, not HIV (Swart-Kruger and Richter 1997).

In other respects, street children differ little from most poor children (Kilbride *et al.* 2000). Despite engaging in petty crime, most have mainstream moral attitudes (Ennew 1994). There is no clear evidence that life on the streets damages cognitive functioning and most children function well emotionally (Aptekar 1989). Street children's social networks are often well developed and friendships are a very important source of emotional support (Ennew 1994). Nutritional status may even be higher than among other poor children (Baker 1996).

Children's opinions about street life vary. In Indonesia, children tend to regard life on the streets favourably (Beazley 2000b); most street children in Brazil do not (Raffaelli *et al.* 2001). Those who left home in search of adventure or to join friends are usually more positive than those who ran to the streets under duress (Raffaelli *et al.* 2001). Research with Bangladeshi street children revealed a number of key concerns: physical and verbal abuse from adults and the police; problems with work (dislike of their job, inability to find better

Table 7.3 *Approaches to street children*

	Correctional model	*Rehabilitative model*	*Outreach strategies*	*Preventive approach*
View of street children	Public nuisance and risk to security	Damaged	Oppressed	On streets because of social and economic forces
Objective	Protect public and deter children from life of crime	Rehabilitation to re-enter mainstream society	Empowerment based on Paulo Freire's model of education	Ameliorate the situations that lead children to the streets
Method	Juvenile justice and detention in jail/institutions	Humane programmes of drug detoxification, education and provision of family-like environment	Outreach education including practical and political skills provided to children on the streets; support groups	No simple solutions. Target unemployment, poor housing, etc. and campaign for children's rights
Actors	Government, police	Churches, NGOs	Street teachers and support groups funded by NGOs and church groups	NGOs, coalitions of street children lobbying government

Source: Based on Rizzini and Lusk (1995).

work); lack of a guardian (to help them find work; stand up for their rights; enable girls to marry); poor income; lack of access to education; and anxieties about the future (West 1999). Older children often become dissatisfied, losing sympathy and income once they no longer appear 'cute' (Aptekar 1989; Beazley 2000a).

Table 7.3 outlines four broad approaches to the 'problem' of street children. None has been entirely successful, even on its own terms. The correctional approach removes children temporarily from streets, but as nothing is done to address the problems that put them there, most later return. Rehabilitative models can help only a tiny proportion. Furthermore, removing children from the streets causes problems, including breaking up support networks (Ennew 1994). Children used to freedom often find institutions difficult, and many return to the streets (Raffaelli *et al.* 2001). Outreach strategies recognise children's right to be in the streets, but the dangers in being 'of the streets' (Scheper-Hughes and Hoffman 1998). While sometimes beneficial, they can be paternalistic ('befriending' children who have their own friends; providing a 'father figure' because female parenting is assumed inadequate (Ennew 1994)). Outreach also sometimes treats street children as inevitably apart from mainstream society, providing education appropriate to street life, but not helping them leave the streets (de Moura 2002). Preventative programmes aim to address the reasons why children go to the streets in the first place, but even these risk overemphasising and perhaps stigmatising the role of families (de Moura 2002).

Sexual exploitation of children

In the late 1980s and early 1990s, sensationalist reports of escalating numbers of children engaged in commercial sex work attracted the attention of international NGOs. The statistics were often absurdly inflated, calculated on the basis of unwarranted assumptions rather than empirical research (Rosemberg and Andrade 1999). In 1986, for example, it was estimated that there were five million prostitutes in Brazil aged 12 to 14 – roughly half the age cohort. Yet again, the Third World poor were stigmatised en masse, through the implication that most poor families were willingly selling their children into prostitution (Rosemberg and Andrade 1999). Yet the stereotype of the child prostitute remained extremely narrow: a young HIV-positive girl rescued from a brothel, having been kidnapped, trafficked, forced into debt bondage or tricked into prostitution. The conflation of all sex workers under the age of 18 into the single category 'child prostitute' obscures differences of age and gender, as well as the different ways that children and youth enter prostitution, and the varied settings in which they work.

Children work as prostitutes in a variety of settings: on the streets, in brothels, hotels and bars. A minority are forcibly recruited, others fulfil parental expectations, but many work for an income for themselves, as well as supporting their families. Many young Malawian women in their mid-teens to early twenties, for instance, work as bar girls, ostensibly employed to serve drinks. In compensation for low pay, they are provided with a rent-free room behind the bar to use for commercial sex. By engaging in sex work, generally without their parents' knowledge, these young women with basic primary education can earn as much as a beginning graduate (Kishindo 1995). A very different situation prevails in urban Thailand where some slum communities are economically dependent on the prostitution of (often pre-pubescent) girls and boys. Children are introduced to prostitution as young as three, moving to penetrative intercourse after a few years, and by their early teens many are pimping for other children. Many of their clients are wealthy Westerners, some of whom develop long-lasting relationships with the children (Montgomery 2001).

Child prostitution is widely condemned as child sex abuse. Both commercial sex and particularly sex with children are subject to highly censorious moral discourses in the West and elsewhere, and perpetrators of child sex abuse are perhaps the most despised in society. Yet discussion of sexual exploitation of children should not be limited to transactional sex. In Karnataka, India, for instance, the *devedasi* practice involves the dedication of daughters of the lower castes to a village deity before puberty, following which they are made available to priests and other men for sex (Verma and Saraswathi 2002). It is also noteworthy that many young people under 18 engage in consensual sex, which may or may not be transactional. In southern Africa, many girls are described as 'prostitutes' because they have sexual relationships with boys or older men for which they receive material benefits – food, presents, etc. The limits of the term 'sexual exploitation' are not easily defined.

Beyond moral outrage, international concern about the sexual exploitation of children has focused in part on the dangers to which children are exposed. Pre-pubescent children's bodies are easily damaged by penetrative sex with adults (Montgomery 2001). For young children prostitution is also emotionally dangerous. Evidence from the West, while it may not be directly applicable in other cultural contexts, points to life-long psychological and social problems among survivors of child sex abuse (Montgomery 2001). Children and adolescents are also at greater risk of contracting disease than adults (Willis and Levy 2002). Between 50 and 90 per cent of children removed from brothels in parts of Asia are HIV-positive. Adolescent girls have a 1 per cent risk of acquiring HIV during each act of unprotected sex with an infected partner, but are less able than older women to negotiate the use of condoms (Willis and Levy 2002). If they have another STI the risk quadruples. Children's lack of access to sexual health services increases their vulnerability. Many with HIV contract tuberculosis. Cervical cancer is associated with low age at first intercourse and multiple sexual partners (Willis and Levy 2002). Without contraception, sexually active adolescents also stand a 90 per cent chance of becoming pregnant within a year, and both pregnancy and abortion carry higher risks than for adults (Willis and Levy 2002).

In seeking to explain child prostitution, media attention has focused on the demand for child prostitutes, particularly from Western 'sex tourists' (in practice, most clients are local men). The sex industry worldwide is estimated to generate $20 billion or more a year, $5 billion of which is attributed to child prostitution (Willis and Levy 2002), and often involves organised crime. Explanations of why men desire sex with children range from paedophilia to the belief that AIDS can be cured by sex with a virgin, or at least that young girls are less likely to transmit HIV.

The existence of child prostitution depends, however, not only on demand but also supply. Although some children are recruited by force, usually children and their families are involved in the decision to engage in prostitution. Attitudes differ between societies. Thailand, for instance, is not alone in having a long history of families selling children to men to pay off debts (Montgomery 2001). Although Western attitudes to child prostitution increasingly prevail among the middle classes, in some northern Thai communities sex work is seen as risky rather than immoral. Parents are happy if their daughters contribute to family income, and less concerned about how they earn money (Sacks 1997).

Child prostitution needs to be understood in relation to the context of poverty in which it takes place. In poor Thai slums, people have no jobs, no land and no welfare provision. Although children do not like prostitution, they have few alternatives if their families are to survive. 'Prostitution, for the children themselves, is not an issue of morality versus immorality but of turning a socially unacceptable form of earning money into a way of fulfilling their familial obligations' (Montgomery 2001: 157).

Whether child prostitution is associated with force or extreme poverty, it is frequently argued that children cannot give informed consent to sex with an adult. A typical view is put forward by Willis and Levy (2002: 1417) 'children [who they define as anyone under 18] cannot consent to being prostituted because, in addition to child prostitution being illegal and a violation of human rights conventions, children do not have the requisite capacity to make such decisions'. Research with young prostitutes, however, reveals that most claim to make the decision themselves. Clearly, children's 'choice' to pursue prostitution is not based on full information, and their options are limited by structures beyond their control. Children are also influenced by their parents. They may have few if any viable alternatives, and do not see prostitution as a 'good' way of life, but, at the same time, children do not passively comply with their circumstances, but struggle to take control of decisions (Montgomery 2001).

Policy responses to child prostitution advocated by international NGOs reflect a Western iconography of child sex. The focus in Thailand has been on men who exploit children and, because Western condemnation of local sexual practices would be considered culturally insensitive, Western tourists are targeted (Montgomery 2001). This narrow focus ignores the vast majority of child prostitution and deflects attention from the reasons why children offer themselves/are offered as prostitutes. While children and their families see prostitution as a lesser evil than succumbing to poverty, the structural economic inequalities that drive children to prostitution receive inordinately less condemnation than is reserved for those who pay for sex with children (Montgomery 2001).

Furthermore, prosecution of clients through the criminal justice system can harm children. Forced to testify against their exploiters, children may be permanently stigmatised as former child prostitutes, and hence unemployable (Montgomery 2001). Equally, threatening families with prison for prostituting their children is unlikely to help children, and may make them more isolated and reluctant to ask for help (Montgomery 2000).

In focusing on clients, children themselves have been neglected. Although prostitutes earn more money than they would in other jobs, ultimately they remain poor (Montgomery 2001). Alternatives are needed, both for those who have not yet entered the sex industry, and for those who would like to leave it. An NGO programme in northern Thailand provided information to six communities about the health risks involved in child sex work, particularly HIV, and offered alternatives in the form of vocational training and loan schemes. The proportion of female school leavers entering the commercial sex industry dropped from 80 per cent to 2 per cent (Sacks 1997).

The stereotype of child prostitutes as innocent victims of an enormous crime has made policy responses to real child prostitution problematic. Few children see themselves as defined by prostitution (Montgomery 2001). Those who work irregularly from home or have run away of their own accord may be viewed as undeserving of sympathy (Montgomery 2001). 'So central is the desexualisation of children to their (covertly sexualised) representation that the "knowing" child is rendered culpable' (Burman 1995b: 128). Emphasising innocence and denying children's agency also risks undermining children's capacities to look after themselves. The media often represent child prostitutes as invariably HIV-positive and therefore bound to die. By denying that they have a future, governments feel absolved of the need to make financial provision, or to offer rehabilitation (Montgomery 2001).

Children affected by AIDS

Whereas street children and child prostitution have long histories, but have only recently attracted international attention, AIDS has only begun to affect children's lives over the past two decades, and the numbers affected are dramatically escalating, particularly in sub-Saharan Africa where the adult HIV-prevalence rate exceeds 30 per cent in several countries. While attention has focused on orphans, children are affected not only by parental death, but also by the strains that AIDS imposes on their families, communities and even public services such as education and health care (Foster and Williamson 2000), especially where poverty is already a problem. When parents or other immediate family members become ill, there is often a decline in the resources available to children, both material and emotional. Children may be expected to substitute for the sick person, carrying out domestic chores, childcare, agricultural work, or sometimes earning an income. They may also take on caring roles, cleaning, feeding and administering medicine to a relative with AIDS, either within their immediate household, or sometimes resident elsewhere (Robson 2000; Robson and Ansell 2000). School attendance by children from AIDS affected households is often poor and many drop out because of work commitments or lack of financial resources (Figure 7.1). Some children have been described as 'virtual orphans' because, although their parents are still alive, they are too ill to care for them: such children may be cared for by other relatives.

Globally more than 13 million children are estimated to have lost one or both parents to HIV (USAID/UNICEF/UNAIDS 2002). Although there is nothing new about orphanhood, the scale of the AIDS pandemic, and attendant problems related specifically to AIDS, exacerbate the difficulties children face. In rich countries, usually no

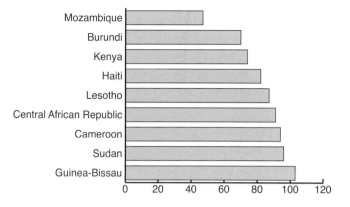

Figure 7.1 *Orphans' school attendance as a percentage of non-orphans' attendance.*

Source: Based on UNDP (2003).

more than 1 per cent of children are orphaned; in the Third World, before AIDS, the figure was about 2 per cent (Hunter and Williamson 2000). By 2001, in 10 African countries more than 15 per cent of children were orphans, as many as three-quarters of these due to HIV (USAID/UNICEF/UNAIDS 2002) (Table 7.4).

Distinguishing 'AIDS orphans' from other orphans in service delivery is inappropriate and often stigmatises those affected. However, AIDS orphans are distinct insofar as they are much more likely to become double orphans. In many cases, fathers predecease mothers, yet most estimates of orphan numbers refer only to maternal orphans. Those who have lost fathers to AIDS are less easily enumerated, but can suffer serious economic consequences (Foster and Williamson 2000). Parental death affects children differently at different ages. Orphan estimates generally relate only to children under 15, excluding older adolescent girls who may be vulnerable to sexual and economic exploitation on the loss of a parent. Inclusion of 15–17-year-olds can inflate the total number by 25–35 per cent (Foster and Williamson 2000).

Following the death of a parent, many children's domestic circumstances are disrupted. Children may continue to live with a surviving parent, but, in some patrilineal societies, a widow who refuses to marry her brother-in-law may be expelled from the home with or without her children. If the death of a parent leaves children with no resident carer, children may be fostered by relatives resident in another household, either locally or at a distance (Box 7.5).

Table 7.4 Estimated orphan numbers in selected African countries

Country	Year	Orphans as % of all children	AIDS orphans as % of orphans	Total orphans		Maternal		Paternal		Double	
				Total	AIDS-related	Total	AIDS-related	Total	AIDS-related	Total	AIDS-related
Botswana	1990	5.9	3.0	34,000	1,000	14,000	<100	23,000	1,000	2,000	<100
	1995	8.3	33.7	52,000	18,000	19,000	7,000	37,000	13,000	5,000	3,000
	2001	15.1	70.5	98,000	69,000	69,000	58,000	91,000	69,000	62,000	61,000
Lesotho	1990	10.6	2.9	73,000	<100	31,000	<100	49,000	<100	8,000	<100
	1995	10.3	5.5	77,000	4,000	31,000	1,000	52,000	4,000	7,000	1,000
	2001	17.0	53.5	137,000	73,000	66,000	38,000	108,000	63,000	37,000	32,000
Malawi	1990	11.8	5.7	524,000	30,000	233,000	11,000	346,000	23,000	55,000	6,000
	1995	14.2	24.6	664,000	163,000	305,000	78,000	442,000	115,000	83,000	41,000
	2001	17.5	49.9	937,000	468,000	506,000	282,000	624,000	315,000	194,000	159,000
Uganda	1990	12.2	17.4	1,015,000	177,000	437,000	72,000	700,000	138,000	122,000	44,000
	1995	14.9	42.4	1,456,000	617,000	720,000	341,000	1,019,000	450,000	282,000	211,000
	2001	14.6	51.1	1,731,000	884,000	902,000	517,000	1,144,000	581,000	315,000	257,000

Source: Based on USAID/UNICEF/UNAIDS (2002).

Fostering of children is common in Africa, not only in enforced circumstances. Research in Tanzania found fewer than a quarter of foster children were orphans (Foster and Williamson 2000). In places severely affected by AIDS, however, willing foster families are increasingly difficult to find. In Zambia, at least one orphan is cared for in over 35 per cent of households (Fussell and Greene 2002). While, traditionally, orphans in patrilineal societies were cared for by paternal uncles and their families, today many such families are too stressed to accept more children, and increasing numbers are cared for by maternal grandparents (Nyambedha *et al.* 2003). Foster families, themselves, can be affected by AIDS, further disrupting the lives of orphans. Although it is seldom favoured, sometimes siblings are split to share the burden of care. Furthermore the presence of orphans in a household can adversely affect the other resident children (Foster and Williamson 2000).

Box 7.5

Children's AIDS-related migration in Lesotho and Malawi

Like most southern African countries, Lesotho and Malawi are characterised by high levels of family dispersal: a legacy of long histories of labour migration. Because of this, strategies for dealing with AIDS commonly involve young people in migration between households of the extended family. Such migration takes place both locally and over longer distances, and often involves moving from urban to rural areas. Families use children's migration as a strategy to meet children's needs, and also to employ children's capacities to address the needs of people in other households. Children may be sent to be cared for elsewhere, because their immediate household is unable to provide adequate care; some are sent to care for sick relatives; and others go away to work (paid or unpaid) to help support themselves or others. The migration patterns that children engage in are highly complex, with many children undertaking multiple migrations in response to changing situations.

AIDS-affected children face a number of difficulties relating to migration, which are often exacerbated by AIDS. Some children have difficulty fitting into new families where they may feel discriminated against and/or have to work hard. Children also have to accommodate to new communities, which may involve having to make new social contacts, attend a new school and (especially in rural areas) learn to undertake unfamiliar forms of work. Most find ways of coping with migration, but these sometimes involve adopting harmful behaviours such as smoking or drinking in order to fit in.

Sources: Ansell and Young (2004); Young and Ansell (2003a; 2003b); Ansell and van Blerk (2004).

Children orphaned by AIDS face a range of difficulties, some of which are shared by other AIDS-affected children. Children whose close relatives are sick or die may experience psychosocial distress, including anxiety and depression (Pivnick and Villegas 2000). The stigma attached to AIDS as a sexually transmitted infection can also cause problems for children (Webb 1997). Many children experience a worsening economic situation, and disruption to their education. Research in several African countries has shown that orphans are more likely than other children to be below the appropriate grade in school (Bicego *et al.* 2003), due to infrequent attendance, poor performance, temporary withdrawal or being put back a class due to migration.

Some children are discriminated against by foster families, or have to share the attention and resources of a grandparent caring for several families of orphans. Children who are fostered are more likely to drop out of school, engage in paid and unpaid work and suffer abuse (UNAIDS 1999; Urassa *et al.* 1997). Where children receive inadequate economic and emotional support, they may also have a heightened vulnerability to HIV infection (Foster and Williamson 2000).

Despite its shortcomings, family or community-based care is generally considered preferable to institutional care for AIDS-affected children, not least because institutional care on the scale required would be prohibitively expensive. Nonetheless, there are situations where no foster family is available, or where arrangements break down. In some sub-Saharan African countries, significant

---

Box 7.6

Strategies to assist children affected by AIDS

1 Strengthen and support the capacity of families to protect and care for their children;
2 Mobilise and strengthen community-based responses;
3 Strengthen the capacity of children and young people to meet their own needs;
4 Ensure that governments protect the most vulnerable children and provide essential services;
5 Create an enabling environment for affected children and families.

Source: Hunter and Williamson (2000).

numbers of children live in sibling-headed households with little adult involvement in their lives. Other children, particularly boys, seek survival on the city streets. (Girls are more likely to find foster homes, serving as unpaid domestic helpers (Boyden 1991).)

Given that, even after the infection rate begins to fall, HIV/AIDS will continue to affect children for at least another 30 years (Foster and Williamson 2000), governments and NGOs are eager to find ways to support those affected. Approaches range from increasing access to anti-retroviral therapies in order to keep carers alive and healthy for longer (Ireland and Webb 2001) to feeding schemes for infants and income-generation schemes for older orphans (Box 7.6).

Children with disabilities

It is estimated that 85 per cent of children with disabilities inhabit Third World countries (Milaat *et al.* 2001). Children experience disabilities associated with sensory, mobility, mental and intellectual impairments (see Figure 7.2); they may be born with these impairments or acquire them later through illness, accident or abuse. The high incidence of disabilities in the Third World relates to poverty and the environmental factors that underlie high levels of illness and accidents; the prevalence of unattended home births; and the fact that selective abortion, which has reduced the number of children born with impairments in the West, is seldom available.

Attitudes towards people with disabilities vary historically and culturally, and partly reflect whether a person is considered able to perform useful tasks in a socially acceptable way (Kabzems and Chimedza 2002). In much of Africa, disabled children were historically seen as burdens to their families and communities, believed to be 'unnaturally conceived, bewitched and, therefore, neither fully human, nor a part of the community' (Kabzems and Chimedza 2002: 150). In ChiShona, spoken in Zimbabwe, nouns describing people with disabilities usually begin with the prefix 'chi', which is normally reserved for objects, and used for humans only in pejorative contexts. Negative attitudes to disabled children persist, particularly among fathers, as disability is associated with maternal wrongdoing, witchcraft, evil spirits, punishment or tests by God (Kabzems and Chimedza 2002). Many families in India similarly attribute the birth of a disabled child to past misdemeanours, often viewing it as somewhat unnatural, and families are affected by guilt,

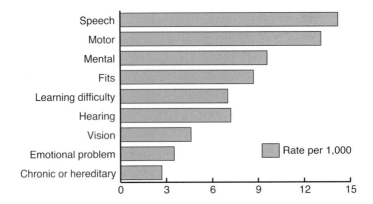

Figure 7.2 *Frequency of disabling conditions among children aged 0–15 in eastern Jeddah area, Saudi Arabia.*

Source: Based on Milaat *et al.* (2001).

fear and stigma (Alur 2001). Because of stigma, some families keep children with intellectual disabilities hidden (Dawson *et al.* 2003).

Provision for disabled people in Third World societies is also affected by Western understandings of disability. The dominant Western model of disability has been a 'medical model', in which disability is seen as synonymous with individual physical (or sometimes intellectual) impairment (Table 7.5), and is understood to require treatment or rehabilitation of the individual. Disability is also viewed as a 'personal tragedy', to be addressed by charities and voluntary organisations rather than a mainstream government concern (Alur 2001). This approach is favoured by donors who like to be seen as giving to the 'less fortunate' (Kabzems and Chimedza 2002).

Recently a 'social model' of disability has gained dominance among disability activists and, increasingly, policy-makers in the West. Disability is understood as a socially constructed form of exclusion: society disables people with physical and intellectual impairments through discrimination or failure to make reasonable adjustments for their needs (Gatrell 2002). People with similar impairments in different societies experience differing levels and forms of disability. This model advocates the removal of barriers, which may be physical, attitudinal, institutional or legal, to the participation of disabled people. Disability is an issue that impinges not only on health and social welfare – it also needs to be understood in relation to politics, economics, development and human rights (Miles 1996).

Table 7.5 *Medical, charitable and social models of disability*

Model	Cause of disability	'Remedy'	Key actors
Medical/ charitable	Individual impairment, 'personal tragedy'	Medical treatment and 'rehabilitation'	NGOs, charities, health service
Social	Society's attitudes towards people with impairments	Political, human rights approach	Disabled people's organisations, all institutions

In Puerto Rico, for instance, physically impaired children experience more limited mobility than their peers in the mainland USA. Inaccessible public buildings, lack of equipment and inaccessible transportation, combined with busy, steeply inclined roads that often lack pavements make it difficult for children with disabilities to leave their homes to attend school, and houses themselves are difficult to move around in a wheelchair (Gannotti and Handwerker 2002). Similarly, in rural India, children with epilepsy have been shown to have more limited social interaction with their peer groups than could be explained by their neurological impairment (Pal *et al.* 2002). Equally, even relatively mild learning difficulties are likely to predispose children to school failure and disable them far more in societies where progression through school is contingent on academic success at each level (Christianson *et al.* 2002).

Although the CRC recognises disability as a ground for protection against discrimination, in most Third World countries, government provision for children with disabilities is minimal (Holloway *et al.* 1999). Even in Jeddah, Saudi Arabia, only about a third of children receive any service provision (Milaat *et al.* 2001). 'In countries where a large proportion of the people live in poverty, governmental priorities may not emphasize the needs of disabled people' (Nordholm and Lundgren-Lindquist 1999: 516). However, as childhood mortality has declined in some countries, the interest of healthcare planners in disabled children has increased (Christianson *et al.* 2002).

Since 1976, the WHO has been committed to the promotion of community-based rehabilitation (CBR) in Third World countries, to assist people with disabilities to live as independent lives as possible through the involvement of all levels of society from the national to the local community (Nordholm and Lundgren-Lindquist 1999). Although CBR is considered to be adaptable to different cultural contexts, its emphasis on minimising the impairments of individual disabled people and promoting independence is accused of Western bias (Miles 1996). The use of the term 'rehabilitation' suggests a medical model of disability, although many CBR programmes now prioritise access to education and employment rather than medical rehabilitation (Miles 1996).

The prevalence of medical and charitable models of disability among governments, donors and NGOs has had marked impacts on education for disabled children. It is estimated that only 2 per cent of disabled children in the Third World attend school. Many countries do not recognise disabled children's right to an education. In Lesotho, children with disabilities were long considered the responsibility of the Ministry of Health, not the Ministry of Education. Although Zimbabwe introduced universal primary education in 1987, children with disabilities were not included (Kabzems and Chimedza 2002). Eighteen countries have laws precluding severely disabled children from access to public education (Lansdown 2001a). Even when the right of a disabled child to attend school is recognised, they may not be provided with appropriate education, and legal loopholes persist in phrases such as 'funds permitting' (Kabzems and Chimedza 2002).

In both India and southern Africa, the only education specifically for disabled children was provided in a few special schools run by churches or NGOs (Kabzems and Chimedza 2002). Most disabled children who could attend school were in mainstream classrooms, though without specific provision. When 'inclusive' education became popular internationally in the 1980s, the challenge was to find ways of addressing children's needs within mainstream settings (Alur 2001). In Lesotho, for instance, a programme was developed that sought to change teachers' attitudes, and provide knowledge and instructional materials to enable them to deal with children with a broad range of needs within large mainstream classrooms (Kabzems and Chimedza 2002). However, where structural adjustment has made school fees and transport excessively costly for many families, disabled children are usually the first to stay at home (Kabzems and Chimedza 2002).

Policy and practice towards children with disabilities is contentious, with four key interest groups pursuing distinct agendas (Box 7.7). Disability professionals usually manage government and NGO programmes, and have access to financial resources and institutional power. Disabled people's organisations (DPOs) often seek to promote social rather than medical models of disability. These are

Box 7.7

Deaf children in Zimbabwe – integration or segregation?

Before Zimbabwe gained independence in 1980, the only educational provision for deaf students took the form of residential special schools. The 1987 Education Act advocated integration of children with disabilities, emphasising commonalities rather than differences between children, and seeking to promote a view of disabled people as normal. This move towards integration, while popular among parents of deaf children and disability professionals, has been opposed by deaf activists in Zimbabwe who argue that placing a deaf child in a mainstream class does not make education 'normal' for the child. Deaf students are often viewed with suspicion by hearing classmates, have considerable difficulties communicating, and do not become fully integrated. But more significantly, integrated education threatens the existence of deaf culture in Zimbabwe. Adults who attended schools for deaf children in Zimbabwe have their own culture, expressed through Zimbabwe Sign Language. Without special schools, there are few opportunities in a country such as Zimbabwe for deaf children of hearing parents to meet other deaf children or to learn sign language.

Source: Chimedza (1998).

generally dominated by adult men, and show little interest in children's issues, particularly the concerns of children with profound or multiple disabilities (Miles 1996). In southern Africa, parents' pressure groups also lobby for influence. These are usually dominated by mothers, many of whom are abandoned by their partners when a child is born or diagnosed with a disability (Miles 1996). DPOs often conflict with parents, accusing them of neglecting, overprotecting or discriminating against disabled children. Lastly, disabled children themselves have minimal influence (Kabzems and Chimedza 2002).

Children in institutions

Many of the groups of CEDC discussed above also fall within UNICEF's category of 'children without primary care givers'. The vast majority of children in the Third World who cannot live with their own families are cared for within extended family or community contexts, but there are about six to eight million children worldwide who live in some form of institutional care, the majority of them in the Third World (Tolfree 1995). Although few non-Western societies employed institutional care in pre-colonial times, institutionalisation is now seen as the normal response to meeting the needs of children in difficult circumstances among many organisations and governments in Third World countries (Tolfree 1995).

Institutional residential care may be defined as 'a group living arrangement for children in which care is provided by remunerated adults who would not be regarded as traditional carers within the wider society' (Tolfree 1995: 6). Institutions vary greatly, ranging from small church or community-based homes, to the internationally funded 'SOS children's villages', to penal institutions. They have been used as a response to family problems (parental death, abandonment, neglect, domestic violence), disabilities, financial problems, separation due to armed conflict or emergencies, and conduct perceived as threatening to society (UNICEF 2003b). In Latin America, for instance, many children are placed in institutions because their parents are considered too poor to care for them. Chile has laws relating to children in 'irregular situations' including delinquency, abandonment and situations where children are deemed in 'physical or moral danger'. Argentina has similar 'laws on abnormal situations' which include being expelled from school, spending time on the streets, experiencing violence at home and delinquency (UNICEF 2003b). In such situations, the State is required to take coercive action both to control conduct and protect children, but no distinction is made between delinquency and abandonment, and many children are deprived of their freedom without a court hearing.

The use of institutions as a means of caring for separated children has fallen out of favour in many parts of the world in recent years, for a number of reasons, ranging from instances of child abuse to inadequate attention to children's emotional needs. The CRC states that the family is the 'natural environment' for children, and considers placement in institutional care an arrangement of last resort (United Nations 1989). Institutionalisation has been accused of operating as a form of 'social exclusion', unnecessarily depriving children of freedom (UNICEF 2003b: vii). Children cared for in institutions may be disadvantaged, first by the difficult situation that led to them being separated from their families; and subsequently by

Plate 7.1 SOS Children's Village, Lesotho.
Source: Lorraine van Blerk.

an institutional environment that usually fails to meet all their needs and fails to prepare them for a future life outside the institution (Tolfree 1995).

While institutions were intended to replace the role of families, the emphasis was on meeting physical needs, and children's need to be loved, cared about and feel a sense of belonging were often overlooked. These needs are more difficult to meet in an institution than in a family environment. The deinstitutionalisation of children in the West was prompted in part by Bowlby's (1966/1951) 'attachment theories' which claimed young children needed a close and stable relationship with one adult. Bowlby's ideas are problematic, particularly in cultural contexts where parenting is usually shared and multiple attachment relationships are normal, but cross-cultural evidence suggests all children have emotional needs that are inadequately met in institutional situations. Infants in institutional care may also receive insufficient individual stimulation for language and motor development, although the long-term consequences of separation and institutionalisation are difficult to establish (Tolfree 1995).

Poverty is increasing in many places, and AIDS is depriving many children of their parents, while urbanisation is eroding some of the networks that traditionally facilitated alternative parenting within the extended family or community. It seems likely, therefore, that the number of candidates for institutional care will continue to rise. Simultaneously, welfare spending is falling under structural adjustment programmes, making it more difficult to provide satisfactory institutional care (Tolfree 1995). Most children in institutional care could be supported by their own families at much lower cost if preventive strategies were employed that ensured the food-security and other basic survival needs of families were met (Tolfree 1995). Equally, substitute family care is generally a better and cheaper solution than institutionalisation for children who cannot remain within their families. If organisations saw children as subjects with rights, rather than objects of charity, they might be more likely to consider alternative forms of provision (Tolfree 1995). Under pressure from the CRC, Guaymallen Province in Argentina is promoting deinstitutionalisation based on principles that include children's participation in decisions affecting them, and recognition of the capacity of poor families to care for and protect their children (UNICEF 2003b).

Equally, it is important to recognise that there is no single ideal alternative to institutional care. Children have differing needs that relate to, among other things, their age, personal characteristics and cultural background. Children are not passive recipients of care, but active agents, and some find ways of improving their own situations within institutions, individually or with their peers. Furthermore, some institutions, however unsatisfactory from a Western perspective, achieve satisfactory outcomes, whereas the inappropriateness of some Western forms of substitute family care may be overlooked (Tolfree 1995).

Conclusions

A small minority of children are deemed to be living in especially difficult circumstances for a number of reasons, often including the fact that they are not living in the stable family environment that Western notions of childhood consider appropriate. There is commonly an assumption that such children are abandoned by their families, yet often the root problem is poverty. There is also a strong tendency to cast CEDC as innocent victims of their circumstances, downplaying their own agency.

Recognition of children's agency is valuable for several reasons. The emphasis on the value of stability in the development and well-being of children is increasingly replaced with an understanding that children engage with their environments in a dynamic way, making choices, defining their own roles and managing crises on the basis of their own interpretations of the changing world around them (Boyden 2000). While young children, in particular, are usually innocent insofar as they cannot be blamed for their situation, the notion of childhood innocence becomes problematic when it is upheld as a virtue and children who take responsibility for themselves are disqualified from sympathy on the basis of being insufficiently childlike. It is also, perhaps, the assumption that CEDC are victims that explains why these children are frequently talked about, but seldom talked to (Montgomery 2001).

Equally problematic in discussions of CEDC is their homogenisation under particular labels, especially the inclusion of all aged under 18 in the category 'child'. What are especially difficult circumstances for an infant may be less so (and certainly very different) for a 17-year-old.

Key ideas

- In contexts of extreme poverty and vulnerability, some children and their families employ coping strategies that place children in especially difficult circumstances.
- CEDC are widely portrayed as innocent victims of poverty, neglect or violence.
- NGOs often downplay the agency of children deciding to participate in prostitution, soldiering or living on the streets.
- Emphasising children's innocence and denying their agency risks undermining their capacities to look after themselves.
- Approaches to CEDC have seldom involved consulting and working with young people themselves.

Discussion questions

- How should societies treat children who have committed crimes?
- How might NGOs mobilise Western support for CEDC without employing images of childhood innocence?
- Do programmes for high-profile CEDC, such as street children and child prostitutes, merely divert funding from the majority of children living in poverty?
- Is there ever a place for institutional care for CEDC?

Further resources

Books

Boyden, J. and Gibbs, S. (1997) *Children of war: responses to psycho-social distress in Cambodia*, Geneva: UNRISD – examines children's responses to a war situation.

Kilbride, P., Suda, C. and Njeru, E. (2000) *Street children in Kenya: voices of children in search of a childhood*, London: Bergin and Garvey – an anthropological study of street children's lives.

Montgomery, H. (2001) *Modern Babylon? Prostituting children in Thailand*, Oxford: Berghahn Books – an anthropological case study of child prostitution in an urban Thai community.

Panter-Brick, C. and Smith, M.T. (eds) (2000) *Abandoned children*, Cambridge: Cambridge University Press – contains chapters based on research with children in a number of situations, including street children.

Singhal, A. and Howard, S.W. (eds) (2003) *The children of Africa confront AIDS*, Athens: Ohio University Press – chapters based on research relating to interventions for children and youth affected by AIDS.

Tolfree, D. (1995) *Roofs and roots: the care of separated children in the developing world*, Aldershot, UK: Arena (for Save the Children) – reports on a Save the Children study of residential institutions and alternative forms of provision for children.

Journals

There are no specific journals focusing on children in especially difficult circumstances. The most useful source of articles is *Childhood*. See, in particular, the 1996 special issue on street and working children.

Websites

Anti-Slavery International http://www.antislavery.org/ publishes news and information about child trafficking and forced labour.

Save the Children http://www.savethechildren.org.uk/ has a section on exploitation and protection.

SOS Children's Villages run institutions for children around the world. Information is available at http://www.sos-childrensvillages.org/.

UNAIDS http://www.unaids.org/ has an extensive range of online publications on children affected by AIDS.

UNICEF http://www.unicef.org/ offers information and online publications under the heading 'child protection'.

8 Rights, participation, activism and power

Key themes

- Children, rights and citizenship
- The UN Convention on the Rights of the Child
- Realising rights to provision and protection
- Participation
- Activism and power.

Chapters 4 to 7 have considered aspects of young people's lives, and raised a number of problems that young people face. One way in which governments and NGOs have sought to relieve such problems is through recourse to a discourse of rights, expressed most comprehensively in the United Nations Convention on the Rights of the Child. The CRC encompasses rights to protection, provision and, most controversially, participation. Although children are social actors and may take on considerable responsibility, they are often silenced and rendered invisible by the attitudes and practices of adult society (Roche 1999). Many claim to feel powerless and excluded. Participation is seen as a way of enhancing children's control over their own lives. However, mobilising around children's rights is unlikely to benefit the majority of children unless it is accompanied by wider political changes that ameliorate the situations in which they live.

Children, rights and citizenship

The discourse of rights has been prominent in development circles in recent years, as bearers of 'rights' are seen as able to make claims with dignity and independence, unlike people with 'needs' who must beg for charity. However, although the League of Nations accepted the Declaration of the Rights of the Child in 1923, the idea that children might be considered to have rights has remained controversial. This controversy arises partly from contradictions in the application of Western discourses of childhood, autonomy and the family.

There are two distinct ways of thinking about rights in relation to children. The first has roots in liberal Western philosophy that sees individuals as entitled to lead their lives as they see fit, provided this does not impinge on the freedoms of others. Individuals are deemed capable of making rational autonomous decisions, and to be the persons best placed to judge their own interests. Children have not traditionally been accorded rights, as they are not understood to possess the necessary rationality, insight or capacity to act in their own interests. The growing view that children are competent social agents has inspired calls to extend to children rights to freedom and self-determination (Archard 1993). The second argument for children's rights rests on the contrary understanding of children as unable to act in their own interests: children are considered to need rights which guarantee them certain forms of treatment such as minimum standards of health care, education, freedom from violence and cruelty. These rights demand no action on the part of children themselves, but require others to secure the appropriate conditions (Archard 1993). These two distinct concepts underlie demands for very different sets of children's rights. While the former view might, for instance, support children's right to work, the latter would require children's protection from economic exploitation.

Some commentators remain uncomfortable with children's rights on either account. 'The language of rights . . . sees society principally as a contractual association of independent, autonomous, self-interested individuals governed by certain rules or principles' (Archard 1993: 88). Some prefer to see children's place in society governed by a moral economy based on affection and care. This is closely associated with the Western view that children belong in the private sphere, where such a moral economy is believed to hold sway.

Box 8.1

Child liberation versus the caretaker thesis

Child liberationists

Prescription:
- Children should have all the rights adults possess – to vote, work, own property, have sex.
- 'Status crimes' (those that are only criminalised when performed by children, such as purchasing alcohol) should be abolished.

Reasons:
- Discriminating against children is equivalent to discriminating against any other social group.
- Age is an arbitrary way of determining who has what rights: it is not directly related to competence – many nine-year-olds have reasoning and moral capacities comparable with many adults.
- Children's incompetence is an ideological construct used to perpetuate their dependence on adults: because children are presumed incapable they are denied opportunities to demonstrate (or even practise) their capabilities.

Problems:
- Very young children do lack the competencies to exercise many 'adult' rights, and are therefore excluded by this very individualist conception of rights.
- Allowing children to make mistakes may harm them.

Caretaker view

Prescription:
- Adults (family or state) should act paternalistically on children's behalf, making the choices they would make for themselves, were they adults.

Reasons:
- Children lack the cognitive capacity and experience to make intelligent decisions, and lack emotional consistency.
- Only if others act in their best interests when they are children will they be able to act in their own best interests as adults.

Problems:
- Adults cannot know what decisions children would make, were they adults.
- 'Best interests' are judged on the basis of familial ideology – treated either as common-sense or based on (often misguided) professional expertise.
- By making decisions on behalf of children, adults shape children's futures, diminishing their capacity to construct their own lives.
- Denying children the opportunity to make mistakes may harm them in the future.

Sources: Archard (1993); Roche (1999); Burman (1995a).

Giving children rights is seen as a state intrusion into the jurisdiction of the family that weakens parental authority.

Archard (1993) counterposes two perspectives on children's rights: the 'child liberation' and 'caretaker' views (Box 8.1). Both are problematic, and most people take an intermediary position: there are certain aspects of life in which children are best placed to determine their own course of action, and children should be permitted to make mistakes, but not mistakes that would seriously jeopardise their future well-being. It might be considered desirable to bring up children such as to maximise their options in adulthood, though it could also be argued that all children need to acquire cultural values as a starting point. The limits of the role of caretaker are thus never clear-cut. Furthermore, it need not be assumed that autonomy is a precondition for an individual's exercise of rights. The liberal concept of rights fails to recognise that all people are in some senses interdependent (Roche 1999), and that 'all rights . . . require resources and state action' (de Waal 2002: 4).

There are countless areas of life in which children may be permitted levels of protection or autonomy, and societies usually accord or remove different rights as children grow older and their needs and abilities change. Age is now commonly used to determine whether a person is allowed to leave school, earn money, marry or face criminal prosecution. Although age is a far from clear-cut indicator of competence, it is difficult to define rights on any alternative basis (Archard 1993).

The nature and extent of the rights accorded to children of different ages are subject to compromise. People in different social and cultural settings, with different concepts of childhood, have different ideas. This conflicts with the conventional Western view that rights are universal – applying to all people on the same basis regardless of culture or individual characteristics. It also raises the question of who should determine children's rights – family, community, state or an international body?

The UN Convention on the Rights of the Child

The CRC, adopted by the UN General Assembly in November 1989 and in force from September 1990, was the most rapidly ratified international convention ever, ratified within eight years by all the

---

Box 8.2

Four principles of the CRC

- *Non-discrimination* (article 2) between children;
- *The best interests of the child* (article 3) as a primary consideration in decisions by state authorities;
- *The right to life, survival and development* (article 6) ensured to the maximum extent possible and including physical health and mental, emotional, cognitive, social and cultural development;
- *The views of the child* (article 12) should be taken seriously in judicial and administrative procedures affecting them.

Source: Office of the High Commission for Human Rights (1993).

UN's 189 member countries except Somalia and the USA. It is 'the first globally-binding treaty protecting children's civil, political, economic, social and cultural rights in both peacetime and armed conflict' (Van Bueren 1996: 27). Although there had been earlier declarations, these were not legal instruments, and most other human rights legislation excludes children.

The CRC reflects a compromise between the liberation and caretaker views. It incorporates three forms of rights: provision (of, for instance, health care and education), protection (from, for instance, violence and neglect) and participation (in decisions affecting their lives). While, as in earlier declarations of children's rights, the 'child's best interests' (Box 8.2) are upheld as paramount, participation rights are introduced for the first time and have proven the most controversial (John 1996). Although children are recognised as complete individuals, with identities distinct from their parents, they are not considered able to exercise rights themselves, hence the rights are translated into duties and responsibilities towards children.

An effect of the CRC is to legitimise the concept of children's rights (Van Bueren 1996). The CRC requires that children be made aware of their rights, and where rights appear in school curricula popular awareness is likely to increase. Since 1989, the CRC has been widely employed by researchers and activists to justify a range of causes, although arguably not always to the real benefit of the children who most need support.

Critiques of the CRC

The most common criticism of the CRC is its Eurocentrism, both in the concept of children's rights it embodies, and in the detail of its prescriptions. The CRC is held to have been drafted by Western-oriented groups who failed to recognise cultural and economic differences in the Third World (Boyden 1990). Rights discourse is by definition generalised and universalised, and assumes an abstract subject, existing outside of gender and cultural inequalities (Burman 1995a). Yet in practice, the subject of the CRC is arguably a child of Western imagination.

Two distinct sets of critiques have emerged. The first relates to the emphasis of the CRC on children's autonomy. The participation articles, in particular, are said to assume socialisation into a Western society that places a higher value on individuality and freedom of self-expression than many other societies (Griesel *et al.* 2002). Anthropologists are uneasy with the presumption of independent rights-bearing individuals, rather than social personhood embedded in larger social units – extended families, clans and villages (Scheper-Hughes and Sargent 1998). Repeated references are made to the roles of 'parents', but not extended families (Boyden 1990). Some suspect a subtext: individual bearers of rights organised in nuclear families are more convenient to global capitalism (Scheper-Hughes and Sargent 1998).

The second set of critiques centres on the representation of children as in need of protection. Ennew (1995) argues the CRC is based on a Western child, believed to belong inside (the home, family and society). Children outside these domains (street children and child soldiers, for instance) become 'outside' childhood, and their needs are not addressed by the Convention. Many of the rights enshrined in the CRC, such as freedom from work, are simply unattainable by the poor, and render children living in poverty 'abnormal' (Boyden 1990). Yet:

> the industrialized North is attempting to impose its own definitions and approaches on the rest of the world, without recognizing the significance, let alone the validity, of other cultures, whilst continuing to enforce the neo-liberal economic policies that are exacerbating the problems of poverty and social deprivation.
>
> (Mayo 2001: 282)

Mayo describes this as 'structural hypocrisy'. Irrespective of whether children should be accorded rights, those inscribed in the CRC are not necessarily the most helpful for poor children.

Alternative sets of rights have been advocated. The African Charter on the Rights and Welfare of the Child (ACRWC), adopted by the Organisation for African Unity in 1990, is very similar to the CRC. Children are granted the right to express an opinion, but, unlike the CRC, there is no requirement for children's opinions to be taken into account. The African Charter also specifies children's responsibilities, including 'to work for the cohesion of the family, to respect his [sic] parents, superiors and elders at all times and to assist them in case of need' (African Union 1999, Article 31). Interestingly, African states were much slower to sign the ACRWC than the CRC. This may suggest their motives for signing the CRC related more to international status than a true commitment to children's rights (de Waal 2002). Alternatively, it may be that the ACRWC differs so little from the CRC as to be considered superfluous. More radical rights are occasionally suggested. Ennew (1995), for instance, advocates children's right to work, to have their own (extrafamilial) support systems respected, not to be labelled and to control their own sexuality.

Another problematic issue with the CRC is its application to all under the age of 18. Although the CRC appears to be geared around the needs and interests of middle childhood, it has begun to entrench an arbitrary cut-off point of 18 that does not correspond with the gradual way adult status is often acquired in Third World societies (de Waal 2002). Furthermore, in applying only to those under 18, the CRC assumes all over 18 have full adult rights. In many societies, young people over 18 continue to feel discriminated against. The Braga Youth Action Plan (United Nations 1998) calls for a youth rights charter: a youth-friendly compendium of existing youth rights emanating from UN resolutions and international conferences.

Realising rights to provision and protection

Institutionally, realising the rights inscribed in the CRC is primarily the responsibility of governments. Governments have responded in various ways: many have established ministries with responsibility for children (and also youth, although the two are often distinct). Some have developed plans, such as Sri Lanka's five-year plan for children, which specifies targets to be met (Van Bueren 1996). Uniquely among international conventions, however, the CRC requires action not only from national governments, but specifies international responsibilities. While multilateral organisations such as the IMF have not acceded to the CRC, most international donors have and are expected to abide by it both domestically and in their development interventions (de Waal 2002). Many NGOs, too, are enthusiasts for children's rights.

Compliance with the CRC is monitored by the Committee on the Rights of the Child, an independent international body. Governments are required to make their laws compatible with the CRC. This is difficult in many African countries, where 'customary law' or interpretations of Islamic law have jurisdiction at the local level: local courts are often reluctant to integrate international law with local customs (Temba and de Waal 2002). South Africa, however, took this into account in drawing up its constitution (Temba and de Waal 2002).

The CRC can only be pleaded directly in a national court of law if its provisions are directly incorporated into national law. Provision rights in particular (survival, health, nutrition and education) demand resources from governments and are difficult to inscribe in legislation where countries lack economic resources. The West has seldom been as keen to promote social and economic rights in poor countries as civil and political rights, and many of the CRC provision articles are limited 'to the maximum extent possible'. This lack of enforceability is the reason the USA cites for not signing the Convention (de Waal 2002). Furthermore, children commonly lack legal capacity to pursue their own rights. In Guatemala, where children are entitled to go to court to defend their rights to protection, this has not ended their abuse (Box 8.3). However:

> [d]ocuments should not be dismissed because their authority is symbolic rather than statutory. Acknowledging the legitimacy of a group's claim to rights is itself part of the process of empowerment; rights are levers which the empowered group must pick up and put to work.
>
> (Franklin 1992, cited in John 1996: 13)

The international pressure exerted on Guatemala as a signatory to the CRC led to proceedings against police officers for the murder of street children (Van Bueren 1996).

Box 8.3

Claiming protection rights in Guatemala City

Although in theory street children in Guatemala City have recourse to the courts to defend their rights, they seldom take the police to court despite routine abuses including physical assault and even murder. Research found that most street children understood the concept of human rights and that they could file a charge (*denuncia*) if their rights were infringed, but problems were encountered at each stage in the process of asserting their rights:

Stage 1: Naming (defining incidents as abuses of rights) – many incidents occurred so frequently that the children did not recognise them as abuse.

Stage 2: Blaming (locating the blame for an injury) – many children blamed themselves, feeling that they had left home so it must be their fault if the police treated them badly; many did not wish to see themselves as victims – to claim their rights would be to acknowledge they needed help from mainstream society.

Stage 3: Claiming (seeking a remedy from the appropriate authority) – most would not do this because they feared reprisals or simply did not believe it would work. The few children who did file *denuncias* were those who were less marginalised, maintaining links with mainstream society.

Source: Godoy (1999).

Participation

Although participation is only directly mentioned in the CRC in relation to children with disabilities, there are four articles that outline children's rights to participate (Box 8.4). These rights may be restricted only where they conflict with laws that are necessary for the protection of national security, public safety, order, health or morals, or the rights, reputations and freedoms of others (United Nations 1989). The CRC conceives of participation in two senses – possibilities for children to engage with the world around them and opportunities to have a voice in more formal decision-making processes (Stephens 1994).

While the CRC does not give children the right to autonomy or allow them to control all decisions that affect themselves, 'it does introduce a radical and profound challenge to traditional attitudes, which assume that children should be seen and not heard' (Lansdown 2001b: 2). Parents, professionals and politicians are obliged to enable

---

Box 8.4

Excerpts from the participation articles of the CRC

Article 12   States Parties shall assure to the child who is capable of forming his or her own views the right to express those views freely in all matters affecting the child, the views of the child being given due weight in accordance with the age and maturity of the child.

Article 13   The child shall have the right to freedom of expression; this right shall include freedom to seek, receive and impart information and ideas of all kinds, regardless of frontiers, either orally, in writing or in print, in the form of art, or through any other medium of the child's choice.

Article 14   States Parties shall respect the right of the child to freedom of thought, conscience and religion . . . States Parties shall respect the rights and duties of the parents and, where applicable, legal guardians, to provide direction to the child in the exercise of his or her right in a manner consistent with the evolving capacities of the child.

Article 15   States Parties recognise the rights of the child to freedom of association and to freedom of peaceful assembly.

Source: United Nations (1989).

---

and encourage children to contribute their views on matters that affect them in the family, school, their communities and nationally. All children, irrespective of age, are deemed capable of expressing a view, but the amount of weight given to children's views is expected to reflect their understanding of the particular issues involved.

Although children's participation has only recently gained international support, the participation of youth has been an aspiration of the UN since 1965. 1985 was designated International Youth Year: Participation, Development, Peace (United Nations 1996). The UN defines youth participation as comprising four components: economic participation (work and development); political participation (decision-making processes); social participation (community involvement); and cultural participation (the arts, cultural values and expression) (Lansdown 2002). The World Programme of Action for Youth emphasises the contribution that youth can make and the unique perspectives they offer, particularly through youth organisations (United Nations 1996).

Arguments for participation

The case for children's participation is still far from universally accepted. Children have seldom been expected to contribute their knowledge and ideas to decision-making processes, it being assumed that adults can speak on their behalf. Children are considered too inarticulate, or to express views that are inconsistent with their experiences, or simply not to tell the truth. But all of these characteristics are also true of adults, and adults are often poor interpreters of children's lives (Boyden 2003).

Children are often more competent than is assumed. Even small children are used to making decisions about friendships, and negotiating rules of games, and many have wider responsibilities (Lansdown 2001b). It is sometimes argued that expecting children to participate detracts from their right to a childhood free from adult concerns. This very Western notion of childhood neglects the fact that children are influenced by the same economic and social forces as adults (Matthews *et al.* 1999). Many children struggle individually against situations of poverty, but are denied the means to effect real change. For this and other reasons, children often want to become involved (Save the Children 2002b).

Not only do children often have the interest and capacity to participate in decision-making, their involvement brings wide-ranging benefits (Box 8.5). As key stakeholders with direct and relevant experience in relation to matters affecting their lives, children can contribute to better decisions (Save the Children 2002b). Adults trying to act in children's best interests have often made mistakes. The failed policy of institutionalising street children would have been abandoned sooner had children been asked their opinions (Lansdown 2001b). Giving young people information to allow them to make their own choices also helps them to protect themselves (Lansdown 2002).

Meaningful participation constitutes an education for active citizenship: something that is not magically attained on reaching the legal voting age (Chawla 2002b). Through participation children learn to become effective in challenging the sources of their own exploitation and to develop their own agendas for transformation (Mayo 2001). Thus participation is empowering of children, both in the present and in the future.

It is also worth stating that children's participation can be encouraged for reasons that are unhelpful to children. Viewing

---

Box 8.5

Outcomes of involving children in development

Outcomes for children   More positive sense of self; increased sense of competence; greater sensitivity to the perspectives and needs of others; greater tolerance and sense of fairness; increased understanding of democratic values and behaviours; preparation for lifelong pattern of participation; new social networks; new skills; enjoyment.

Outcomes for organisations that serve children   Programme and policy development that is sensitive to children's priorities; the establishment of processes for participation; increased commitment to children's rights; innovation.

Outcomes for children's communities   Public education regarding children's rights; more positive public attitudes and relationships to children; increased social capital; improved quality of life.

Source: Chawla (2001).

---

children as actors, rather than objects of development, is not an argument for seeing them as a resource rather than as citizens (Johnson 1996). Furthermore, participation is not a 'free good' and does not automatically lend validity to a project (UNICEF 2003d).

Facilitating participation

As UNICEF (2003d: 3) points out, '[i]n truth, children have always participated in life: in the home, in school, in work, in communities, in wars'. Young people make decisions and take actions whether adults are involved or not. However, legal, economic and social climates can by altered such that children are better able to exercise autonomous or collective action in their own interests. The purpose behind the CRC participation articles is 'to optimise [children's] opportunities for meaningful participation' (UNICEF 2003d: 4). This can mean providing opportunities to contribute to decisions that affect children as individuals – about, for instance, their education and health, or which parent or relative they will live with, as well as decisions and activities that affect other children and adults. Participation may take place in any environment – schools, workplaces, health services, local and national government – or in their own organisations – clubs, unions, parliaments, etc., at all levels from the family to the international arena (Lansdown 2002). Young people's participation has often been facilitated by NGOs including development NGOs, service organisations for young people and youth organisations, and involved activities including project design, management, monitoring and evaluation, community activities, policy-making, campaigning and lobbying, conference participation and research (Lansdown 2001b; Mayo 2001).

Roger Hart's (1992) ladder of participation (Figure 8.1) differentiates between ways of involving children. The lower three rungs of the ladder are not viewed as participation and are widely considered undesirable. Manipulation describes situations where adults use children's voices to carry their own messages. Decoration might, for instance, involve children wearing t-shirts promoting a cause they know nothing about. The upper five rungs represent different levels of participation. Hart does not suggest the highest rungs are 'better', but that children's own characteristics and specific contexts will require different degrees of participation.

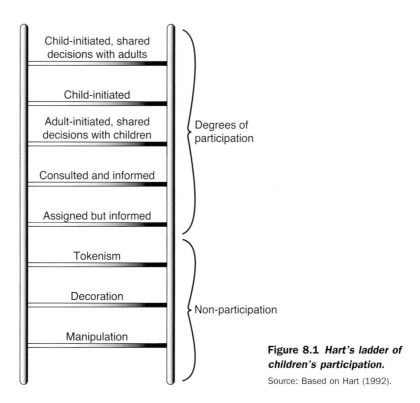

Figure 8.1 *Hart's ladder of children's participation.*

Source: Based on Hart (1992).

Much participatory work with young people begins with a process of finding out about their lives, the problems they face and their priorities for change. Christensen and Prout (2002) identify four ways in which research processes have involved children: the child as object (known, rather than knowing, information is sought about rather than from them); as subject (having feelings and opinions of their own); as a social actor (taking part in, changing and changed by the social and cultural world they inhabit); and as a participant and co-researcher (on the basis that children are active participants in societal life). Past research into children's lives tended to rely on accounts from parents and caregivers. Progressively children have been allowed more autonomy, initially acting as respondents in their own right (Hutchby and Moran-Ellis 1998), and more recently participating in all aspects of the research process, from identifying the issues to be researched, choosing the methods, conducting the research, interpreting results and disseminating findings (Boyden and Ennew 1997)

Research with children requires careful consideration of processes (Box 8.6), although rigid ethical guidelines are inappropriate as children and their situations differ (Christensen and Prout 2002). Innovative methods have often been employed (Box 8.7), in recognition of children's competencies and, more significantly, their marginalisation in adult society (Punch 2002a). To avoid the abuse of power, it may be preferable to avoid conducting research in places that children perceive as adult-controlled, and to use group techniques to alter the power balance. In view of children's widely differing individual preferences and competencies, there is no ideal method for research with children, and 'the suitability of particular methods depends as much on the research context as on the research subject's stage in the life course' (Punch 2002a: 338).

Mayo (2001) points out that the real challenge is often not so much to enable children to speak as to ensure their voices will be heard. Many advocate 'giving voice' to young people in accounts of research, but this is not easy where what they say conflicts with established understandings of phenomena, and it is seldom appropriate to take all that children (or adults) say at face value. Children are increasingly given opportunities to present their own viewpoints in public arenas, including local and international conferences. While potentially valuable, such participation can be merely tokenistic. The official youth delegation to the Earth Summit in Rio, 1992, were promised an hour to present their statement.

Box 8.6

Questions for researchers

Purpose of research   What is it for? Whose interests does it serve? Will it help children? Is it necessary?

Costs and potential benefits   What contribution is needed from children? Are there risks or costs? Can these be reduced?

Privacy and confidentiality   How will confidentiality be preserved? What about revelations of abuse?

Selection, inclusion and exclusion   Are disadvantaged groups represented? Are allowances needed to help them participate? Single-sex groups? Will work or school be impeded?

Funding   How is the research funded? Are children's/guardians' expenses covered? Should children be paid/rewarded?

Review/revision of aims and methods   Have children helped plan/comment on these? Have they been piloted?

Information for children and guardians   Are all details of aims, methods and possible outcomes made clear?

Consent   How easily can they opt out? Is parental consent needed?

Dissemination   Are children involved? Who will it be disseminated to?

Impact on children   How does the research impact on children more widely? Are children represented in positive ways?

Source: Adapted from Alderson (1995).

When they arrived they were told they would have only 10 minutes. After two minutes the TV cameras were turned off and the reporters in the pressroom could not hear (John 1996).

Young people's participation extends beyond the provision of knowledge and viewpoints that help others act in their interests. Young people actively participate in organisations and in activities intended to improve the lives of themselves and others (see Boxes 8.8 and 8.9). Children's participation at community level is said to be developing faster in some Third World countries than in Western

---

Box 8.7

Methods used in research with young people

- Visual techniques, e.g. drawings, diagrams, maps, storyboards;
- Role-play, drama and songs, usually improvised by children;
- Photo appraisals and video-making;
- Children's writings, including essays, diaries, recall and observations;
- Semi-structured or unstructured interviews.

Source: Wilkinson (2000a).

---

democracies (Hart 1997). In Nepal, over 30,000 children aged 8 to 16 are involved in a new form of children's club, managed, to varying degrees, by children themselves. Like adult organisations in Nepali communities, most have an executive board of seven to nine children including a chairperson, vice-chairperson, secretary and treasurer. Although younger, low caste and ethnic minority children are underrepresented on the boards, many children learn skills and knowledge they would not otherwise acquire, including how organisations function and how to manage them. All members experience democratic decision-making, and design and participate in community development projects (Rajbhandary *et al.* 2001). Young people are also involved in activities at national and international levels.

Facilitating children's participation is beset by various difficulties. First, children participating in research or speaking at conferences are often assumed to be representative of children in general. They are seldom, however, elected representatives, putting forward the views of others. Sometimes children are chosen to embody the diversity of a project's intended beneficiaries, in terms of, for instance, age, gender, ethnicity and disability, but often the more disadvantaged children remain excluded. It is frequently falsely assumed (including by children themselves) that older children can be articulate representatives of younger children (Bourdillon 2000c). There is also a risk of children becoming 'professionalised' speakers, increasingly distanced from the children they purport to represent (Lansdown 2001b).

Second, while children are not powerless, they can be manipulated by adult agendas. Furthermore, if children perceive adults as

---

Box 8.8

Children helping infants in Nigeria

In Afugiri, a densely populated peri-urban community of about 25,000 people in Umuahia, Abia state, Nigeria, the well-equipped and easily accessible primary health care facilities were barely used. Although there were 1,000 children aged 0–11 months, over an 11 month period only six to eight infants were immunised each month, only five to seven women attended antenatal care services each month and only six babies were delivered in the clinic in eight months. But the local secondary school had a child rights club and the 10–16-year-old members decided to do something about the low rates of immunisation. They received one-and-a-half days of training by UNICEF field officers and health ministry officials and then organised discussions about health and mobilised women to take their children for immunisation. They subsequently went from house to house, identifying eligible infants, handed tracking slips to parents and older children who were asked to take the eligible children to the health centre, and traced those who did not show up. An average of 328 infants were immunised in each of the eight months of the project. The health centre provided further services to the mothers: education in safe motherhood, prevention and home management of common illnesses; provision of oral dehydration salts; growth monitoring and teaching of exclusive breastfeeding and complementary feeding. The additional services attracted more women to the centre and monthly attendance rose to over 300 women a month and the number of babies delivered at the facility more than doubled. Many of the schoolchildren followed up those infants who had had their first immunisation at the clinic, ensuring that they had the full three doses of the DPT vaccine. Furthermore, other states in Nigeria were so impressed with the results that they plan to copy the project.

Source: UNICEF (2003d).

---

powerful, they may hesitate to report their genuine views. It is necessary to establish trust and to try to distinguish between normative statements derived from popular discourse and those more closely reflecting children's own feelings and experiences (Johnson 1996). There are also power relations between child participants, and it is necessary to ensure that, for instance, girls and younger participants are not marginalised. On occasions, a dominant child may take over and manipulate both adults and children (Johnson 1996). Furthermore, there has been a tendency to promote leadership skills rather than participation skills, with little benefit to the majority of young people (Lansdown 2002).

Third, children's participation sometimes creates tensions within communities, particularly where adults feel excluded. This was a

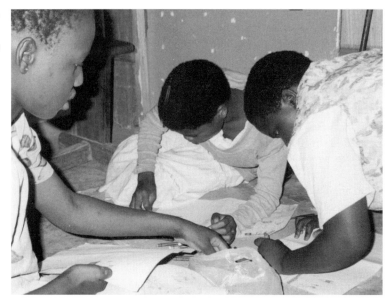

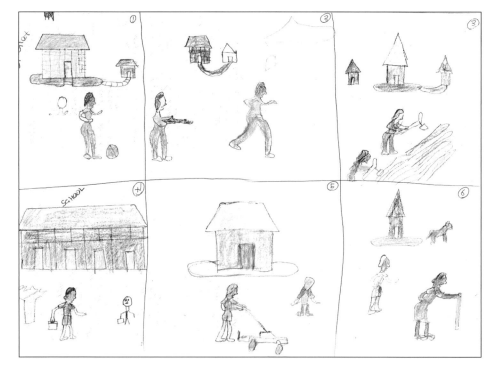

(b) Migration storyboard.

Source: Malawian boy, aged 13.

Plate 8.1 Research using storyboards in Malawi.

common problem with the 'Child-to-Child' approach, developed following the Alma Ata International Conference on Primary Health Care (1978) and the UN Year of the Child (1979), based on the observation that children in Third World countries often take responsibility for younger siblings. Children were given basic health information to share with younger children, peers, families and communities, thereby acting as agents of change. In many societies, however, there is no expectation that adults should listen to children,

---

Box 8.9

Growing Up in Cities

In both Agenda 21, the action plan drawn up at the UN Conference on Environment and Development in Rio in 1992, and the Habitat Agenda, agreed at the UN Conference on Human Settlements, 1996, young people are recognised not only to have particular needs with respect to the environment, but also to be well equipped to make a contribution to human settlement development. Through participation in environmental projects, young people may also acquire a long-term interest in the environment. Effective education for sustainability is said to require education *about* the environment (information); education *in* the environment (time spent out of doors); and education *for* the environment (learning how to act effectively on issues of concern) (Chawla 2002b).

Growing Up in Cities (GUIC) is a UNESCO-funded project that was first undertaken in the 1970s and revived in the 1990s: 10–15-year-olds in poor urban neighbourhoods around the world engage in research, action and dissemination of their ideas. GUIC has five objectives:

- Gaining an understanding of children's environmental interests and needs through participatory research;
- Applying this information to the design of programmes and activities to improve life quality for children and their communities;
- Pressing for effective urban policies for children;
- Organizing public events to draw attention to urban children's rights and needs;
- Increasing the capacity for participatory research and action among academic researchers and the staff of community-based organizations .

(Griesel *et al.* 2002: 84)

One of the cities where GUIC has been adopted with enthusiasm is Johannesburg, South Africa. Children, randomly selected to provide a socio-economic and gender balance, were asked to identify problems in their neighbourhoods and in the wider city that could be addressed by local council officials or within urban planning more generally. Multiple methods were used – drawings, stickers on maps to show good and bad places, transect walks. Problems were ranked and solutions proposed (Figure 8.2) (Swart Kruger and Chawla 2002).

Insufficient and unsafe places to play	
Children are inhibited from using the few parks and play spaces that exist because of bullies, hostile adults, broken equipment, litter and rules prohibiting many forms of play	• Train staff in parks and swimming pools to work with children to reduce harassment and violence • Consider creating protected rooftop play spaces
Pedestrian problems	
Frequent pedestrian accidents involving children	• Repair faulty traffic signals rapidly • Provide mechanisms to slow down traffic in residential areas
Public transport	
This is inefficient and expensive	• Introduce a wider range of public transport • Free or reduced fares for children
Harassment and public safety	
Children are harassed in many ways by both adults and children	• Train police to listen to and assist children and to take threats against them seriously
Waste management and littering	
Children are concerned about the health hazards and are offended by having litter dumped in the places where they play and live	• Prioritise clearing litter and waste in places children occupy
Taverns, *shebeens* and bottle stores	
Children experience verbal and physical abuse from the patrons	• Zone drinking establishments away from residential areas • Enforce regulations against open drinking in public areas

Figure 8.2 *Examples of recommendations to improve conditions for children in metropolitan Johannesburg.*

Source: Swart Kruger and Chawla (2002: 96).

and children seen to be giving advice to parents or elders aroused hostility (Mayo 2001).

Last, if expectations are not made clear from the start, children can come to expect far greater change to result from their participation than can realistically be achieved. When expectations are not met, children may feel disempowered and disillusioned by democratic processes.

Activism and power

While young people should arguably have the right to express their views regardless of the outcome, encouraging participation will not automatically bring substantive improvements in young people's lives. The UN Research Institute for Social Development distinguishes 'systems maintaining' participation, used to make people more receptive to existing policies, from 'systems-transforming' participation, which encourages people to engage in real decision-making, even where this challenges existing authority structures (Chawla 2002a). In few cases has children's participation led to young people seriously challenging authority. Indeed, some forms of participatory development have been associated with neo-liberalism and efforts to appropriate free labour from the poor (Mayo 2001).

Young people have thus far been given little opportunity to participate in areas that really make a difference. 'Much of government policy impacts directly or indirectly on young people's lives, yet it is developed and delivered largely in ignorance of how it will affect the day-to-day lives of young people, their present and future well-being' (Lansdown 2002: 6). While young people may be permitted to participate in politics with a specific children/youth focus, they are not invited to participate in the formulation of policy in the wider social and economic spheres. Even education, despite its profound impacts on young people's lives, has seldom been opened up to participation, although the Committee on the Rights of the Child asks governments to report on children's participation in their school systems. This reflects schools' roles in maintaining social stability (John 1996).

It is issues such as budget allocations and structural adjustment programmes that really impact on young people's lives. These are areas in which Third World people in general (not only the young) are disenfranchised through the influence of the IFIs. So long as structural conditions persist which prevent children's lives from being ameliorated, the right of young people to express a view is arguably irrelevant. Indeed, children's participation may be a low-cost exercise that makes children appear responsible for removing themselves from a situation of poverty that can be addressed only by those with real political power.

If the lives of children and youth are to improve, political change is needed. Yet the widespread rhetorical support for children's issues rests on them being considered apolitical. The speed of ratification of the CRC attests to the unwillingness of governments to be seen as against children. Aid organisations tend to downplay political and adversarial aspects of campaigns, as uncontroversial images are more effective for fundraising. Children's parliaments, similarly, have deliberately been made as non-political as possible. Although usually led by youth, they tend to focus on children's rights, which are less contentious than the demands of youth (Temba and de Waal 2002).

This is not to deny children's agency. While all young people are social actors striving to improve their individual lives, individual acts of survival lack the political salience of collective acts of resistance (Scheper-Hughes and Hoffman 1998). Roche (1999: 478) points out that 'children do resist and challenge adult practices, though not necessarily in obvious or constructive ways. However, the choices available to children, as a relatively powerless group in society, differ from those available to the relatively powerful.' There are limits to children's power, enforced by families, by society and more especially by a global economy that leaves them poor. Children's effective exclusion from public space makes it harder for them to make their views known, and resistance is often symbolic rather than instrumental (Roche 1999). Nonetheless, street children in Brazil have been able to attract both national and international media attention by marching through the streets of the capital protesting about the lack of government action and even government complicity in the murder of street children (Van Bueren 1996). In Colombia, a visit to a high school by a UN activist prompted the student council to organise a campaign against the violence that beset the country. First they collected over 5,000 stories, poems and pieces of art by youth expressing protest; they organised 'peace carnivals' and eventually brought together the 'Children's Movement for Peace' – a national movement which organised a referendum in which three million youth voted for 'rights to survival, peace, family, and freedom from abuse' (Welti 2002: 354–355). As a consequence, peace was a central issue in the 1998 presidential elections, and the turnout of adult voters doubled.

Young Africans have engaged in collective rebelliousness around numerous social and political issues (de Waal 2002). In Africa 'the young are numerous, energetic, and seek alternatives to their plight. Young people do not simply succumb to the domination and repression of authoritarian states; they do not accept their limited life options but actively seek alternatives' (Galperin 2002: 133). On 16 June 1976 thousands of South African children joined a protest in Soweto against the use of Afrikaans in schools, seen as the language of the oppressor. Police opened fire killing 500 people, spurring many youth into more active resistance to Apartheid (Reynolds 1995). Across Africa, the political activists who achieved independence defined themselves as 'youth', to distinguish themselves from the elders who they saw as collaborators with colonial rule. In the post-Independence era, however, gerontocratic political structures have re-emerged, many of the 'youth' leaders of 40 years ago having hung on to power. Today student unions are again active in both universities and secondary schools in Africa, and university students in particular have been at the forefront of recent radical political change in several countries (de Waal 2002).

Youth activism also has a long history in India. Thousands of students discontinued their studies to join the Quit India movement in 1942, and young people were involved in the fight for social justice following independence (Verma and Saraswathi 2002). Political parties established power bases on university campuses in the 1950s, and in the 1960s and 1970s students were active opponents of government, using protests, demonstrations, strikes and boycotts to achieve their goals. The All-Assamese Students' Union, which fought for use of the Assamese language and reservation of key jobs for state residents, formed the first student government in the country (Verma and Saraswathi 2002).

Youth activism is seldom welcomed by governments. In Southeast Asia, '[y]outh are . . . encouraged to exhibit greater civic consciousness. But when they do express their views about social justice and change, they are labeled as "dissidents" by the state' (Santa Maria 2002: 203). Where young people are perceived as potentially hostile, governments seek to restrain their influence:

they fear that giving children civil and (still more so) political rights
may lead to adversarial relations with adults who have duties to care
for children. In countries where young people have strong traditions of
political militancy, governments are unwilling to cede them greater
participation rights.

(Temba and de Waal 2002: 219)

In some contexts, young people's political engagement is particularly
problematic. In the 1990s, Indian youth politics were inspired by
regionalism and religious fundamentalism and characterised by
militancy and violence (Verma and Saraswathi 2002). The Taliban
('Islamic student') government of Afghanistan, comprising both
youth and adult men, was extremely repressive towards women.
In other places, young people have lost interest in politics: despite
the 1989 student protests in Tienanmen Square, few young people
in China today are politically active (Brown and Larson 2002).

It is often argued that young people's peaceful and positive
participation in public life requires formal political rights, including
the right to vote. 'Real change will occur when political leaders have

Plate 8.2 'Student-MP' Nelson Chamisa, Zimbabwe.
Source: Leo Zeilig.

no option but to listen to young people's concerns – because they need young people's votes' (Temba and de Waal 2002: 221). Bosnia-Herzegovina, Brazil, Croatia, Cuba, Iran, Nicaragua, the Philippines, Serbia, Montenegro and Slovenia all allow voting below the age of 18 (Lansdown 2001b). In some cases, young people are encouraged to be more actively involved in formal politics. As Youth MP for Western Uganda, Okwir Rabwoni held one of several seats in the Ugandan parliament reserved for those under 30. This enabled him to promote the engagement of young Ugandans in national politics by establishing village youth committees and a Youth Entrepreneurial Scheme (Rabwoni 2002). Yet even where political parties make space for youth involvement, the outcomes are not invariably favourable towards young people (see Box 8.10).

If teenagers can bring about political change, what of younger children? John (1996) draws on Steve Biko's recipe for black empowerment to suggest that children's empowerment must start with children themselves, and requires responsibility (on the part of children), unity and a broad children's movement. All of these are problematic, though, particularly as children are far from a homogeneous group.

Very young children cannot purposefully exercise power in the political sphere, but have needs which are peculiar to the very young. Newborns cannot agitate for drugs to block mother-to-child transmission of HIV. Under-fives cannot lobby for increased resources to be devoted to immunisation. Even older children have great difficulty challenging from within the practices and processes that exclude them (Matthews *et al.* 1999). A number of commentators have argued that any campaign to realise child rights needs to be led by young people (e.g. de Waal 2002). Yet the issues that rouse youth to action are not the grievances of young children. Youth cannot be expected to champion the rights of the very young to provision and protection and certainly cannot 'participate' on their behalf. While there are strong commonalities between the interests and needs of 15- and 20-year-olds, youth have no greater moral obligation towards, or common cause with, young children than do adults.

There are limits to what children can do alone – as there are with any social group. If there are to be improvements in the conditions in which young people live, concessions must be made by others on the basis of a broad consensus of what is right. Most policies and

Box 8.10

Party politics in Botswana

In Botswana, the voting age was reduced from 21 to 18 years old by a referendum requested by the main opposition party, the Botswana National Party, which had benefited from the votes of disillusioned unemployed youth in 1994. The BNP had historically assigned a significant role to youth and in 1994 a number of graduates of its student revolutionary study groups were elected to urban seats. Following the referendum, youth were increasingly targeted for political mobilisation. At the 1999 elections, however, the BNP lost support, and turnout among 18–20-year-olds was lower than among older adults. Research among young people suggested that only 44 per cent of them saw positive benefits from reducing the voting age. Such benefits included changes in government/policies, more sensitivity/responsiveness to youth needs, more employment for youth, greater youth participation in decision-making and the fact that young people are more educated than their elders (almost 60 per cent of 15–19-year-olds were students in 1991) would mean the introduction of innovative ideas and greater understanding of modernity and its demands. More than half the respondents, however, considered that youth were not interested in politics, were not mature and would therefore not contribute to change, or were insufficiently interested or knowledgeable to respond. Most of the young people (68 per cent) thought it was good to vote, and 60 per cent believed in multi-party democracy.

There has been something of a generational change within the main political parties as power has gradually been wrested from the founding veterans. Youth activists tend to develop their interest in politics at high school, and to become politically active at tertiary level. All the main parties have youth wings, and although some members are not strictly youth, youth wings are more effective within the parties than in the past. Women however, are underrepresented, (more so in youth wings than the wider parties), unwilling to tolerate the personal attacks and character assassination youth politics entails. Within the ruling Botswana Democratic Party that has held power since independence, youth have demanded and achieved a limit to the president's term of office, more democratic procedures within the party, and a more competitive system for primary elections.

Source: Selolwane (2003).

programmes that really affect young people affect adults too, if not always in the same ways, hence solutions need to be negotiated, taking the effects on everyone into account. Mobilisation around the situation of young people needs to be seen as a political rather than a charitable cause, although the ill-being of children does not pose an immediate threat and is therefore likely to be perceived as of lesser urgency by many governments (Temba and de Waal 2002).

If political action, at least on behalf of young children, must remain the responsibility of adults, it is essential to ensure that policies are responsive to children's needs, where possible either involving them directly or incorporating their views. There remains considerable scope for increasing young people's participation. The Declaration on the Survival, Protection and Development of Children made at the World Summit for Children in 1990 assumed that the political action it required would be by adults, yet it could easily have specified some involvement by children themselves (Van Bueren 1996). Although children were involved in the UN General Assembly Special Session on children, they were not invited to contribute to the outcome document 'A World Fit for Children' (Lansdown 2002). Even child-focused NGOs seldom involve children in deciding priorities at planning and programming levels (Lansdown 2002). There have been calls for the institutionalisation of young people's participation – legislation could require participatory structures in schools, formal mechanisms for dialogue with policy-makers at all levels and a reduced voting age (Lansdown 2002). The Poverty Reduction Strategy Papers that have so far been drafted in Africa treat children as a target group for health and education, not critical actors in the economy, society, security or governance (Save the Children 2002a). There is also considerable scope for a closer interaction between institutional supporters of young people and young activists themselves. 'In Africa today, it is highly significant that the specialist institutions for children – including UNICEF – are disconnected from the most vibrant movements of young people' (de Waal 2002: 12). Although a role remains for adults, there is good reason to make it easier for young people to initiate and pursue actions on their own behalf, which would constitute real empowerment. However, given the chance to participate politically, to make their voices heard in public, young people will not necessarily see themselves as the best placed people to further their own agendas. Seventeen-year-old Eliza Kantardzic from Bosnia-Herzegovina was one of three war-affected children permitted to address the UN Security Council at the Special Session: 'The best thing you can do to help children in war,' she told them, 'is to stop war, to prevent it. And that is something that this Council has the power to do. The real question is – is that power used?' (UNICEF 2003d: 64).

Conclusions

The international discourse on children's rights has focused attention on age-based discrimination, on questions of cultural relativism and particularly on ways to facilitate children's participation in decisions affecting their lives. In many ways this has been positive. Discrimination against children is now discussed in many societies. Governments have begun to reflect on how they should deal with, for instance, historically entrenched sexual exploitation of children, and radical new constitutions in Brazil and South Africa now enshrine human rights and question aspects of customary law (Scheper-Hughes and Sargent 1998). Children have participated in community projects and international conferences. Although children exercise agency even without formal rights, rights are a tool that can be used to alter the balance of power and facilitate agency: 'participation and the language of children's rights presupposes and encourages their agency, the expression of their self-defined needs and interests' (Roche 1999: 484).

There are, however, areas that have been neglected. Much rights-based NGO programming has neglected children's basic needs (Seaman 2000). Young people's concerns focus not just on age-based marginalisation but also other forms of subordination that affect them. More significantly, rights-based discussions often place children outside politics. They are seen as amenable to solutions that focus on their own particular predicaments without having to address the dirty adult world of politics and economics. Children are the ideal focus for NGOs who want to combine Western public support with ease in dealing with Third World governments and IFIs. Yet the image of innocent childhood, removed from the adult world, does great disservice to the real lives and real interests of children. Children are as much part of the economy as adults – they are just as affected by economic and social change, and indeed contribute to it as they struggle to make lives for themselves in often adverse conditions.

Key ideas

- The reasons for according rights to children are varied and not universally accepted.
- The UN Convention on the Rights of the Child is the first legally binding global agreement that all children should have rights to protection, provision and participation.
- Rights to protection and provision are difficult to secure, as they demand resources that many governments lack.
- Support for children's right to participate in decisions affecting them is growing, and considerable attention has been given to how this can best be achieved.
- The children's rights discourse sometimes downplays the need for wider political change to address the needs of young people.
- There is a role for activism by young people in the political sphere – but also for action by adults on behalf of young children.

Discussion questions

- Try to reach a consensus with other people on minimum ages at which young people should be permitted to drive, have sex, drink alcohol, work for a wage, marry, etc? What are the underlying reasons for any disagreements?

- Compare the African Charter on the Rights and Welfare of the Child with the CRC. What do you think African leaders objected to in the CRC, and why?

- Do children have responsibilities as well as rights?

- Is children's participation a distraction from or solution to addressing children's basic needs?

Further resources

Books

Archard, D. (1993) *Children: rights and childhood*, London: Routledge – a philosophical exploration of the arguments for and against according children rights. The arguments are made very clearly but, interestingly, despite the date of the publication, no reference is made to the CRC.

Boyden, J. and Ennew, J. (1997) *Children in focus: a manual for participatory research with children*, Stockholm: Radda Barnen – a training manual aimed at project staff in child-oriented NGOs and research institutions who conduct research with and about children.

de Waal, A. and Argenti, N. (2002) *Young Africa: realising the rights of children and youth*, Trenton, NJ: Africa World Press – essays examining the situation of children and youth in Africa, with particular regard to how their rights are met by international, national and civil society, and the active involvement of young people in forging new futures for themselves.

Hart, R. (1997) *Children's participation: the theory and practice of involving young citizens in community development and environmental care*, London: Earthscan – written as a guide to involving children, particularly in relation to the environment.

Johnson, V., Ivan-Smith, E., Gordon, G., Pridmore, P. and Scott, P. (1998) *Stepping forward: children and young people's participation in the development process*, London: Intermediate Technology Publications – a collection of over 60 short essays examining a wide range of ethical and practical issues surrounding young people's participation in research and NGO projects in the Third World.

Journals

PLA Notes 25 (1996) and 42 (2001) on children's participation published by the International Institute for Environment and Development.

Websites

Human Rights Watch http://www.hrw.org/children/ has information and a list of publications relating to a range of children's rights issues.

Child Rights Information Network http://www.crin.org/ provides a very full list of children's rights related publications from a range of NGOs and organisations.

Peace Child International http://www.peacechild.org/ is an NGO that aims to empower children to take responsibility for peace, human rights and the environment. It also has ECOSOC status at the UN.

GUIC http://www.unesco.org/most/growing.htm is the Growing up in Cities website.

 **Postscript**

This book has examined how interconnected global processes are affecting the lives of young people. The effects are both negative and positive. While there are thousands of young people engaged in conflict, sex work and exploitative labour, whose vulnerability may be attributable to economic processes that entrench poverty, it is also true that, at a global scale, children are more likely to survive into adulthood than in the past and more likely to gain literacy. In international circles, children and youth are also more likely to be listened to and taken seriously, as people with rights and the capacity to act, as well as with needs. Furthermore, children are no longer seen as the exclusive constituency of children's NGOs, but are recognised as intended beneficiaries of wider development interventions. There is, currently, a momentum both to improve the situations in which young people live, and to involve them in improving their own lives.

At the same time, there are dangers ahead. First, the image of childhood that inspires those engaged in implementing actions on behalf of young people in the Third World remains very Western. Children can be harmed by taken-for-granted assumptions concerning, for instance, the value of education, the harm done by work and the expectation that they are, and should remain, innocent. Moreover, although in some spheres awareness is growing that young people are social actors in their own right, there are other areas (notably health and education) where young people are still predominantly viewed as recipients of services, if not objects of interventions.

A second, related, risk is that of homogenising young people. While development and globalisation are producing some patterns of change that impact on young people irrespective of geography, these processes also interact with local realities to produce very different outcomes for young people. It is essential that policy-makers and practitioners, as well as those engaged in research with young people, recognise the diversity of young people, both geographically, and along social fractures such as gender, class, ethnicity, dis/ability and, importantly, age. Many international pronouncements concerning children have grouped everyone aged 0 to 17 in a single category. Where teenagers are shoehorned into conceptual frameworks developed in relation to young children, this can obscure rather than illuminate their needs and capabilities. It is for this, among other reasons, that this book has advocated a cross-fertilisation between those working with 'children' and those concerned with 'youth'.

Third, there is a danger of segregating children and youth from wider society, both in real terms, and in the policy circles in which they are considered. If young people are seen as being apart from the mainstream, their interests are likely to be both peripheralised and depoliticised. While apolitical issues may be attractive to development NGOs, it is necessary for those working with/for children and youth to recognise that small, localised actions, however 'participatory', are only small, localised actions. It is arguably time for a move away from narrowly focusing on small-scale apolitical participatory projects to seek mainstream political change in the interests of children and youth. It is also time for all national and international institutions to take children and youth seriously in the way that the NGO sector increasingly does – to recognise that their actions are not neutral with regard to children and youth, but that children and youth are part of their constituencies who need to be both considered and consulted.

References

Accorsi, S., Fabiana, M., Ferrouese, N., Iriso, R., Lukwiya, M. and Declich, S. (2003) 'The burden of traditional practices, *ebino* and *tea-tea*, on child health in Northern Uganda', *Social Science and Medicine*, 57: 2183–2191.

Addison, T. and Rahman, A. (2001) 'Why is so little spent on educating the poor?', *UNU/WIDER Conference on Growth and Poverty*, Helsinki, 25–26 May.

Admassie, A. (2003) 'Child labour and schooling in the context of a subsistence rural economy: can they be compatible?', *International Journal of Educational Development*, 23: 167–185.

African Union (1999) *African Charter on the Rights and Welfare of the Child*, accessed 07/09/04, http://www.africa-union.org/Official_documents/Treaties_%20Conventions_%20Protocols/A.%20C.%20ON%20THE%20RIGHT%20AND%20WELF%20OF%20CHILD.pdf.

Agha, S. (2000) 'The determinants of infant mortality in Pakistan', *Social Science and Medicine*, 51: 199–208.

Aitken, S.C. (2001) *Geographies of young people: the morally contested spaces of identity*, London: Routledge.

Alderson, P. (1995) *Listening to children: children, ethics and social research*, London: Barnardos.

Ali, M., Emch, M., Tofail, F. and Baqui, A.H. (2001) 'Implications of health care provisions on acute lower respiratory infection mortality in Bangladeshi children', *Social Science and Medicine*, 52: 267–277.

Allotey, P. and Reidpath, D. (2001) 'Establishing the causes of childhood mortality in Ghana: the "spirit child"', *Social Science and Medicine*, 52: 1007–1012.

Al-Samarrai, S. and Zaman, H. (2002) *The changing distribution of public education expenditure in Malawi*, Working Paper 29, Washington: World Bank.

Althusser, L. (1972) 'Ideology and ideological state apparatuses', in Cosin, B.R. (ed.) *Education: structure and society*, Harmondsworth: Penguin, pp. 242–280.

Alur, M. (2001) 'Some cultural and moral implications of inclusive education in India: a personal view', *Journal of Moral Education*, 30: 287–292.

Ambadekar, N., Wahab, S., Zodpey, S. and Khandait, D. (1999) 'Effect of child labour on growth of children', *Public Health*, 113: 303–306.

Ansell, N. (2001) '"Because it's our culture!": (re)negotiating the meaning of *lobola* in southern African secondary schools', *Journal of Southern African Studies*, 27: 697–716.

Ansell, N. (2002a) '"Of course we must be equal, but . . .": imagining gendered futures in two rural southern African secondary schools', *Geoforum*, 33: 179–194.

Ansell, N. (2002b) 'Secondary education reform in Lesotho and Zimbabwe and the needs of rural girls: pronouncements, policy and practice', *Comparative Education*, 38: 91–112.

Ansell, N. (2004) 'Secondary schooling and rural youth transitions in Lesotho and Zimbabwe', *Youth and Society*, in press.

Ansell, N. and van Blerk, L. (2004) 'Children's migration as a household/family strategy: coping with AIDS in southern Africa', *Journal of Southern African Studies*, 30: 673–690.

Ansell, N. and Young, L. (2004) 'Enabling households to support successful migration of AIDS orphans in southern Africa', *AIDS Care*, 16: 3–10.

Aptekar, L. (1989) 'Columbian street children: gamines and chupagruesos', *Adolescence*, 24: 783–794.

Archard, D. (1993) *Children: rights and childhood*, London: Routledge.

Aries, P. (1962) *Centuries of childhood*, New York: Vintage Press.

Bahr, J. and Wehrhahn, R. (1993) 'Life expectancy and infant-mortality in Latin America', *Social Science and Medicine*, 36: 1373–1382.

Baker, R. (1996) 'Methods used in research with street children in Nepal', *Childhood*, 3: 171–193.

Bakilana, A. and de Waal, A. (2002) 'Child survival and development in Africa in the 21st century', in de Waal, A. and Argenti, N. (eds) *Young Africa: realising the rights of children and youth*, Trenton, NJ: Africa World Press, pp. 29–54.

Barrett, H. (1998) 'Protest–despair–detachment: questioning the myth', in Hutchby, I. and Moran-Ellis, J. (eds) *Children and social competence: arenas of action*, London: Falmer, pp. 64–84.

Baume, C., Helitzer, D. and Kachur, S.P. (2000) 'Patterns of care for childhood malaria in Zambia', *Social Science and Medicine*, 51: 1491–1503.

Beazley, H. (2000a) 'Home sweet home? Street children's sites of belonging', in Holloway, S.L. and Valentine, G. (eds) *Children's geographies: playing, living, learning*, London: Routledge, pp. 194–210.

Beazley, H. (2000b) 'Street boys in Yogyakarta: social and spatial exclusion in the public spaces of the city', in Bridge, G. and Watson, S. (eds) *A companion to the city*, London: Blackwell, pp. 472–488.

Beazley, H. (2002) '"Vagrants wearing make-up": negotiating spaces on the streets of Yogyakarta, Indonesia', *Urban Studies*, 39: 1665–1683.

Bequele, A. and Boyden, J. (1988) 'Child labour: problems, policies and programmes', in Bequele, A. and Boyden, J. (eds) *Combatting child labour*, Geneva: International Labour Office.

Bi, P., Tong, S. and Parton, K.A. (2000) 'Family self-medication and antibiotics abuse for children and juveniles in a Chinese city', *Social Science and Medicine*, 50: 1445–1450.

Bicego, G., Rustein, S. and Johnson, K. (2003) 'Dimensions of the emerging orphan crisis in sub-Saharan Africa', *Social Science and Medicine*, 56: 1235–1247.

Black, M. (1993) *Street and working children*, Florence: UNICEF International Child Development Centre.

Blanchet, T. (1996) *Lost innocence, stolen childhoods*, Dhaka: The University Press Limited.

Booth, M. (2002) 'Arab adolescents facing the future: enduring ideals and pressures to change', in Brown, B.B., Larson, R.W. and Saraswathi, T.S. (eds) *The world's youth: adolescence in eight regions of the globe*, Cambridge: Cambridge University Press, pp. 207–242.

Bourdillon, M. (2000a) 'Children at work on tea and coffee estates', in Bourdillon, M. (ed.) *Earning a life: working children in Zimbabwe*, Harare: Weaver Press.

Bourdillon, M. (ed.), (2000b) *Earning a life: working children in Zimbabwe*, Harare: Weaver Press.

Bourdillon, M. (2000c) 'Introduction', in Bourdillon, M. (ed.) *Earning a life: working children in Zimbabwe*, Harare: Weaver Press.

Bowlby, J. (1966/1951) *Maternal care and mental health*, New York: Schocken.

Boyden, J. (1990) 'Childhood and the policy makers: a comparative perspective on the globalization of childhood', in James, A. and Prout, A. (eds) *Constructing and reconstructing childhood*, London: Falmer, pp. 184–215.

Boyden, J. (1991) *Children of the cities*, London: Zed.

Boyden, J. (2000) 'Children and social healing', in Carlson, L., Mackeson-Sandbach, M. and Allen, T. (eds) *Children in extreme situations: proceedings from the 1998 Alistair Berkley Memorial Lecture*, London: LSE Development Studies Institute, pp. 58–81.

Boyden, J. (2003) 'Children under fire: challenging assumptions about children's resilience', *Children, Youth and Environments*, 13: 1, accessed 07/09/04, http://colorado.edu/journals/cye.

Boyden, J. and Ennew, J. (1997) *Children in focus: a manual for participatory research with children*, Stockholm: Radda Barnen.

Boyden, J. and Gibbs, S. (1997) *Children of war: responses to psycho-social distress in Cambodia*, Geneva: UNRISD.

Boyden, J. and Myers, W. (1995) *Exploring alternative approaches to combating child labour: case studies from developing countries*, Florence: UNICEF International Child Development Centre.

Bridges-Palmer, J. (2002) 'Providing education for young Africans', in de Waal, A. and Argenti, N. (eds) *Young Africa: realising the rights of children and youth*, Trenton, NJ: Africa World Press, pp. 89–103.

Brown, B.B. and Larson, R.W. (2002) 'The kaleidoscope of adolescence', in Brown, B.B., Larson, R.W. and Saraswathi, T.S. (eds) *The world's youth: adolescence in eight regions of the globe*, Cambridge: Cambridge University Press, pp. 1–20.

Brown, D.K. (2001) 'Child labour in Latin America: policy and evidence', *The World Economy*, 24: 761–778.

Burman, E. (1992) 'Innocents abroad: Western fantasies of childhood and the iconography of emergencies', *Disasters*, 18: 238–253.

Burman, E. (1995a) 'The abnormal distribution of development: policies for Southern women and children', *Gender, Place and Culture*, 2: 21–36.

Burman, E. (1995b) 'Developing differences: gender, childhood and economic development', *Children and Society*, 9: 121–142.

Camacho, A.Z.V. (1999) 'Family, child labour and migration: child domestic workers in Metro Manila', *Childhood*, 6: 57–73.

Camara, G. (1999) *Posprimaria Comunitaria Rural: El desafío de la relevancia, la pertinencia y la calidad*, Mexico: Consejo Nacional de Fonnento Educativo.

Campbell, C. and Mac Phail, C. (2002) 'Peer education, gender and the development of critical consciousness: participatory HIV prevention by South African youth', *Social Science and Medicine*, 55: 331–345.

Caputo, V. (1995) 'Anthropology's silent "others": a consideration of some conceptual and methodological issues for the study of youth and children's cultures', in Amit-Talai, V. and Wulff, H. (eds) *Youth cultures: a cross-cultural perspective*, London: Routledge, pp. 19–42.

Carlton-Ford, S., Hamill, A. and Houston, P. (2000) 'War and children's mortality', *Childhood*, 7: 401–419.

Chakraborty, S. and Frick, K. (2002) 'Factors influencing private health providers' technical quality of care for acute respiratory infections among under-five children in rural West Bengal', *Social Science and Medicine*, 55: 1579–1587.

Chawla, L. (2001) 'Evaluating children's participation: seeking areas of consensus', *PLA Notes*, 42: 9–13.

Chawla, L. (2002a) 'Cities for human development', in Chawla, L. (ed.) *Growing up in an urbanising world*, London: Earthscan, pp. 15–34.

Chawla, L. (2002b) '"Insight, creativity and thoughts on the environment": integrating children and youth into human settlement development', *Environment and Urbanization*, 14: 11–21.

Chimedza, R. (1998) 'The cultural politics of integrating deaf students in regular schools in Zimbabwe', *Disability and Society*, 13: 493–502.

Chisholm, L., Harrison, C. and Motala, S. (1997) 'Youth policies, programmes and priorities in South Africa: 1990–1995', *International Journal of Educational Development*, 17: 215–225.

Christensen, P. and Prout, A. (2002) 'Working with ethical symmetry in social research with children', *Childhood*, 9: 477–497.

Christianson, A.L., Zwane, M.E., Manga, P., Rosei, E., Vente, A., Downs, D. and Kromberg, J.G.R. (2002) 'Children with intellectual disability in rural South Africa: prevalence and associated disability', *Journal of Intellectual Disability Research*, 46: 179–186.

Colclough, C. (1982) 'The impact of primary schooling on economic development: a review of the evidence', *World Development*, 10: 167–185.

Comaroff, J. and Comaroff, J. (1991) *Of revelation and revolution: Christianity, colonialism, and consciousness in South Africa*, Chicago, IL: University of Chicago Press.

Commonwealth Secretariat (1970) 'Youth and development in Africa: report of the Commonwealth Africa Regional Youth Seminar', Nairobi, 1969.

Connolly, M. and Ennew, J. (1996) 'Introduction: children out of place', *Childhood*, 3: 131–145.

Costello, A., Watson, F. and Woodward, D. (1994) *Human face or human facade? Adjustment and the health of mothers and children*, London: Centre for International Child Health, Institute of Child Health.

CountryWatch (2003) 'CountryWatch data', accessed 27/05/04, http://www.countrywatch.com/.

Dawson, E., Hollins, S., Mukongolwa, M. and Witchalls, A. (2003) 'Including disabled children in Africa', *Journal of Intellectual Disability Research*, 47: 153–154.

de Moura, S.L. (2002) 'The social construction of street children: configuration and implications', *British Journal of Social Work*, 32: 353–367.

de Silva, M.W.A., Wijekoon, A., Hornik, R. and Martines, J. (2001) 'Care seeking in Sri Lanka: one possible explanation for low childhood mortality', *Social Science and Medicine*, 53: 1363–1372.

de Waal, A. (2002) 'Realising child rights in Africa: children, young people and leadership', in de Waal, A. and Argenti, N. (eds) *Young Africa: realising the rights of children and youth*, Trenton, NJ: Africa World Press, pp. 1–28.

Deolalikar, A. (1993) 'Gender differences in the returns to schooling and in school enrolment rates in Indonesia', *Journal of Human Resources*, 28: 899–932.

Desai, S. (1995) 'When are children from large families disadvantaged? Evidence from cross-national surveys', *Population Studies*, 49: 195–210.

Dessy, S.E. (2000) 'A defense of compulsive measures against child labor', *Journal of Development Economics*, 62: 261–275.

Dewey, K. (2003) *Guiding principles for complementary feeding of the breastfed child*, Washington: Pan American Health Organization/ World Health Organization.

DFID (1999) *Helping not hurting: an alternative approach to child labour*, London: DFID.

DFID (2000) *Eliminating world poverty: making globalisation work for the poor: white paper on international development*, London: DFID.

Dore, R. (1976) *The diploma disease: education, qualification and development*, London: George Allen and Unwin.

Dorsey, B. (1975) 'The African secondary school leaver', in Murphree, M. (ed.) *Education, race and employment in Rhodesia*, Salisbury, Rhodesia: Association of Round Tables of Central Africa.

Dorsey, B. (1989) 'Educational development and reform in Zimbabwe', *Comparative Education Review*, 33: 40–58.

Dow, W.H. and Schmeer, K.K. (2003) 'Health insurance and child mortality in Costa Rica', *Social Science and Medicine*, 57: 975–986.

Duryea, S. and Arends-Kuenning, M. (2003) 'School attendance, child labor and local labour market fluctuations in urban Brazil', *World Development*, 31: 1165–1178.

Dyer, C. (2002) 'Management challenges in achieving education for all: South Asian perspectives', in Desai, V. and Potter, R.B. (eds) *The companion to development studies*, London: Arnold, pp. 419–424.

Eide, A.H. and Acuda, S.W. (1997) 'Cultural orientation and use of cannabis and inhalants among secondary school children in Zimbabwe', *Social Science and Medicine*, 45: 1241–1249.

El-Gibaly, O., Ibrahim, B., Mensch, B.S. and Clark, W.H. (2002) 'The decline of female circumcision in Egypt: evidence and interpretation', *Social Science and Medicine*, 54: 205–220.

Ellis, S. (2003) 'Young soldiers and the significance of initiation: some notes from Liberia', *Youth and the politics of generational conflict in Africa*, 24–25 April, Leiden: African Studies Centre.

Ennew, J. (1994) 'Parentless friends: a cross-cultural examination of networks among street children and street youth', in Nestman, F. and Hurrelman, K. (eds) *Social networks and social support in childhood and adolescence*, London: de Gruyter, pp. 409–425.

Ennew, J. (1995) 'Outside childhood: street children's rights', in Franklin, B. (ed.) *The handbook of children's rights: comparative policy and practice*, London: Routledge, pp. 201–215.

Fabrega, H. and Miller, B. (1995) 'Adolescent psychiatry as a product of contemporary Anglo-American society', *Social Science and Medicine*, 40: 881–894.

Falkingham, J. (2000) *From security to uncertainty: the impact of economic change on child welfare in Central Asia*, Florence: UNICEF Innocenti Research Centre.

Farmer, P. (1992) *AIDS and accusation: Haiti and the geography of blame*, Berkeley, CA: University of California Press.

Fawcett, C. (2001) *Latin American youth in transition: a policy paper on youth unemployment in Latin America and the Caribbean*, Washington: Inter-American Development Bank.

Fernea, E.W. (1995) 'Childhood in the Muslim Middle East', in Fernea, E.W. (ed.) *Children in the Muslim Middle East*, Austin, TX: University of Texas Press, pp. 3–16.

Finn, J.L. (2001) 'Text and turbulence: representing adolescence as pathology in the human services', *Childhood*, 8: 167–191.

Folasade, I.B. (2000) 'Environmental factors, situation of women and child mortality in southwestern Nigeria', *Social Science and Medicine*, 51: 1473–1489.

Foster, G. and Williamson, J. (2000) 'A review of current literature of the impact of HIV/AIDS on children in sub-Saharan Africa', *AIDS*, 14: S275-S284.

Frank, A.G. (1967) *Capitalism and underdevelopment in Latin America: historical studies of Chile and Brazil*, New York: Monthly Review Press.

Freire, P. (1972) *Pedagogy of the oppressed*, Harmondsworth: Penguin.

Freire, P. (1985) *The politics of education: culture, power and liberation*, Basingstoke: Macmillan.

Fussell, E. and Greene, M.E. (2002) 'Demographic trends affecting youth around the world', in Brown, B.B., Larson, R.W. and Saraswathi, T.S. (eds) *The world's youth: adolescence in eight regions of the globe*, Cambridge: Cambridge University Press, pp. 21–60.

Gailey, C.W. (1999) 'Rethinking child labor in an age of capitalist restructuring', *Critique of Anthropology*, 19: 115–119.

Galperin, A. (2002) 'Child victims of war in Africa', in de Waal, A. and Argenti, N. (eds) *Young Africa: realising the rights of children and youth*, Trenton, NJ: Africa World Press, pp. 105–121.

Gammeltoft, T. (2002) 'Seeking trust and transcendence: sexual risk-taking among Vietnamese youth', *Social Science and Medicine*, 55: 483–496.

Gannotti, M.E. and Handwerker, W.P. (2002) 'Puerto Rican understandings of child disability: methods for the cultural validation of standardized measures of child health', *Social Science and Medicine*, 55: 2093–2105.

Gatrell, A.C. (2002) *Geographies of health*, Oxford: Blackwell.

Godoy, A.S. (1999) '"Our right is the right to be killed": making rights real on the streets of Guatemala City', *Childhood*, 6: 423–442.

Gould, W. (1993) *People and education in the Third World*, New York: Longman.

Griesel, R.D., Swart-Kruger, J. and Chawla, L. (2002) '"Children in South Africa can make a difference": an assessment of "Growing Up in Cities" in Johannesburg', *Childhood*, 9: 83–100.

Griffin, C. (2001) 'Imagining new narratives of youth: youth research, the "new Europe" and global youth culture', *Childhood*, 8: 147–166.

Hall, T. and Montgomery, H. (2000) 'Home and away: "childhood", "youth" and young people', *Anthropology Today*, 16: 13–15.

Halvorson, S.J. (2003) 'A geography of children's vulnerability: gender, household resources, and water-related disease hazard in northern Pakistan', *The Professional Geographer*, 55: 120–133.

Hardman, C. (2001/1973) 'Can there be an anthropology of children?', *Childhood*, 8: 501–517.

Hart, R. (1992) *Children's participation: from tokenism to citizenship*, Florence: UNICEF.

Hart, R. (1997) *Children's participation: the theory and practice of involving young citizens in community development and environmental care*, London: Earthscan.

Heady, C. (2000) *What is the effect of child labour on learning achievement? Evidence from Ghana*, Florence: UNICEF Innocenti Research Centre.

Helleiner, J. (1999) 'Toward a feminist anthropology of childhood', *Atlantis*, 24.

Heuveline, P. and Goldman, N. (2000) 'A description of child illness and treatment behavior in Guatemala', *Social Science and Medicine*, 50: 345–364.

Heywood, C. (2001) *A history of childhood: children and childhood in the West from medieval to modern times*, Cambridge: Polity.

Hilary, J. (2001) *The wrong model: GATS, trade liberalisation and children's right to health*, London: Save the Children.

Hilary, J., Penrose, A., King, F., Heaton, A. and Wilkinson, J. (2002) *Globalisation and children's rights: what role for the private sector?*, London: Save the Children UK.

Hoiskar, A.H. (2001) 'Underage and under fire: an enquiry into the use of child soldiers 1994–8', *Childhood*, 8: 340–360.

Hollos, M. (2002) 'The cultural construction of childhood: changing conceptions among the Pare of northern Tanzania', *Childhood*, 9: 167–189.

Holloway, S., Lee, L. and McConkey, R. (1999) 'Meeting the training needs of community-based service personnel in Africa through video-based training courses', *Disability and Rehabilitation*, 21: 448–454.

Holloway, S.L. and Valentine, G. (2000) 'Children's geographies and the new social studies of childhood', in Holloway, S.L. and Valentine, G. (eds) *Children's geographies: playing, living, learning*, London: Routledge, pp. 1–26.

Hunter, S. and Williamson, J. (2000) *Children on the brink: strategies to support a generation isolated by HIV/AIDS*, Washington, DC: USAID.

Hutchby, I. and Moran-Ellis, J. (1998) 'Situating children's social competence', in Hutchby, I. and Moran-Ellis, J. (eds) *Children and social competence: arenas of action*, London: Falmer, pp. 7–26.

ICRW (2001) *Global change: international agreements on youth*, Washington, DC: International Center for Research On Women.

Illich, I. (1971) *Deschooling society*, London: Calder and Boyars.

ILO (2002a) *Every child counts: new global estimates on child labour*, Geneva: International Labour Office.

ILO (2002b) 'What is IPEC: IPEC at a glance', International Labour Organisation, accessed 13/07/03, http://www.ilo.org/public/english/standards/ipec/about/implementation/ipec.htm.

ILO (2003) *Wounded childhood: the use of children in armed conflict in Central Africa*, Geneva: International Labour Office.

Inkeles, A. and Smith, D.H. (1974) *Becoming modern*, London: Heinemann.

Ireland, E. and Webb, D. (2001) *No quick fix: a sustained response to HIV/AIDS and children*, London: International Save the Children Alliance.

Iyun, B.F. and Oke, E.A. (2000) 'Ecological and cultural barriers to treatment of childhood diarrhea in riverine areas of Ondo State, Nigeria', *Social Science and Medicine*, 50: 953–964.

Jabry, A. (ed.) (2002) *Children in disasters: after the cameras have gone*, London: Plan UK, http://www.plan-uk.org/action/childrenindisasters.

James, A., Jenks, C. and Prout, A. (1998) *Theorising childhood*, New York: Teachers College Press.

James, A. and Prout, A. (1990) *Constructing and reconstructing childhood*, London: Falmer.

Jeffrey, C., Jeffery, P. and Jeffery, R. (2003) 'When schooling fails: young men, education and low caste politics in north India', *Uncertain transitions: youth in a comparative perspective*, presented at a conference at University of Edinburgh, 27/06/03.

Jejeebhoy, S.J. (1998) 'Adolescent sexual and reproductive behavior: a review of the evidence from India', *Social Science and Medicine*, 46: 1275–1290.

Jenks, C. (1996) *Childhood*, London: Routledge.

John, M. (1996) 'Voicing: research and practice with the "silenced"', in John, M. (ed.) *Children in charge, volume 1: the child's right to a fair hearing*, London: Jessica Kingsley Publishers, pp. 3–24.

Johnson, V. (1996) 'Introduction: starting a dialogue on children's participation', *PLA Notes*, 25: 30–37.

Johnson, V., Hill, J. and Ivan-Smith, E. (1995) *Listening to smaller voices: children in an environment of change*, Chard, Somerset: ActionAid.

Johnson, V. and Ivan-Smith, E. (1998) 'Background to the issues', in Johnson, V., Ivan-Smith, E., Gordon, G., Pridmore, P. and Scott, P. (eds) *Stepping forward: children and young people's participation in the development process*, London: Intermediate Technology Publications, pp. 3–8.

Jok, J.M. (2003) 'Sudan's prolonged civil war and the shifting value(s) of children', *Youth and the politics of generational conflict in Africa*, 24–25 April, Leiden: African Studies Centre.

Jourdan, C. (1995) 'Masta Liu', in Amit-Talai, V. and Wulff, H. (eds) *Youth cultures: a cross-cultural perspective*, London: Routledge, pp. 202–222.

Jubilee Debt Campaign (2002) *Factsheet 1: HIPC*, London: Jubilee Debt Campaign.

Kabeer, N. (2000) 'Inter-generational contracts, demographic transitions and the "quantity–quality" tradeoff: parents, children and investing in the future', *Journal of International Development*, 12: 463–482.

Kabzems, V. and Chimedza, R. (2002) 'Development assistance: disability and education in southern Africa', *Disability and Society*, 17: 147–157.

Kassouf, A.L., McKee, M. and Mossialos, E. (2001) 'Early entrance to the job market and its effect on adult health: evidence from Brazil', *Health Policy and Planning*, 16: 21–28.

Kelly, P.F. (1999) 'Rethinking the "local" in labour markets: the consequences of cultural embeddedness in a Philippine growth zone', *Singapore Journal of Tropical Geography*, 20: 56–75.

Khashan, H. (2003) 'Collective Palestinian frustration and suicide bombings', *Third World Quarterly*, 24: 1049–1067.

Kilbride, P., Suda, C. and Njeru, E. (2000) *Street children in Kenya: voices of children in search of a childhood*, London: Bergin and Garvey.

Kilmister, A. (2000) 'Socialist models of development', in Allen, T. and Thomas, A. (eds) *Poverty and development into the 21st century*, Oxford: Oxford University Press/Open University, pp. 309–324.

Kimane, I. and Mturi, A.J. (2001) *Rapidly assessing children at work in Lesotho*, Government of Lesotho with financial assistance from UNICEF.

King, E.M. and Hill, M.A. (eds) (1993) *Women's education in developing countries: barriers, benefits and policies*, Baltimore, MD: Johns Hopkins University Press.

King, K. (1996) 'Microenterprise: macroeconomic environment: revisiting Kenya's informal (*jua kali*) sector against the background of the formal globalising economy', *International Journal of Educational Development*, 16: 417–426.

Kiros, G.-E. and Hogan, D.P. (2001) 'War, famine and excess child mortality in Africa: the role of parental education', *International Journal of Epidemiology*, 30: 447–455.

Kishindo, P. (1995) 'Sexual behaviour in the face of risk: the case of bar girls in Malawi's major cities', *Health Transition Review*, 5: 153–160.

Kolk, A. and Van Tulder, R. (2002) 'The effectiveness of self-regulation: corporate codes of conduct and child labour', *European Management Journal*, 20: 260–271.

Kovsted, J., Portner, C. and Tarp, F. (1999) *Determinants of child health and mortality in Guinea-Bissau: does health knowledge matter?*, Development Economics Research Group, accessed 07/09/04, http://www.econ.ku.dk/DERG/pub/pub.htm.

Kratli, S. (2001) *Education provision to nomadic pastoralists: a literature review*, Brighton: Institute of Development Studies.

Kuper, J. (2000) 'Children in armed conflict: the law and its uses', in Carlson, L., Mackeson-Sandbach, M. and Allen, T. (eds) *Children in extreme situations: proceedings from the 1998 Alistair Berkley Memorial Lecture*, London: LSE Development Studies Institute, pp. 38–47.

Lachaud, J.-P. (2004) 'Modelling determinants of child mortality and poverty in the Comoros', *Health and Place*, 10: 13–42.

Lampietti, J.A., Poulos, C., Cropper, M.L., Mitiku, H. and Whittington, D. (1999) *Gender and preferences for malaria prevention in Tigray, Ethiopia*, Washington: World Bank Development Research Group/Poverty Reduction and Economic Management Network.

Landau, P.S. (1995) *The realm of the word: language, gender, and Christianity in a southern African kingdom*, Portsmouth, NH: Heineman.

Langmore, D. (1989) *Missionary lives: Papua, 1874–1914*, Honolulu, HI: University of Hawaii Press.

Lansdown, G. (2000) 'The convention: history and impact', in Carlson, L., Mackeson-Sandbach, M. and Allen, T. (eds) *Children in extreme situations: proceedings from the 1998 Alistair Berkley Memorial Lecture*, London: LSE Development Studies Institute, pp. 6–18.

Lansdown, G. (2001a) *Children's rights: a second chance*, London: International Save the Children Alliance.

Lansdown, G. (2001b) *Promoting children's participation in democratic decision-making*, Florence: UNICEF Innocenti Research Centre.

Lansdown, G. (2002) 'Youth participation in decision-making', *Expert meeting on global priorities for youth*, New York: United Nations Youth Unit.

Lauby, J. and Stark, O. (1988) 'Individual migration as a family strategy: young women in the Philippines', *Population Studies*, 42: 473–486.

Lerer, L.B. (1998) 'Who is the rogue? Hunger, death and circumstance in John Mampe Square', in Scheper-Hughes, N. and Sargent, C. (eds) *Small wars: the cultural politics of childhood*, Berkeley, CA: University of California Press, pp. 228–250.

Lewis, D. (1998) 'Development NGOs and the challenge of partnership: changing relations between North and South', *Social Policy and Administration*, 32: 501–512.

Liechty, M. (1995) 'Media, markets and modernization: youth identities and the experience of modernity in Kathmandu, Nepal', in Amit-Talai, V. and Wulff, H. (eds) *Youth cultures: a cross-cultural perspective*, London: Routledge, pp. 166–201.

Lieten, G.K. (2000) 'Children, work and education – I: general parameters', *Economic and Political Weekly*, 35: 2037–2043.

Loughry, M. and Flouri, E. (2001) 'The behavioral and emotional problems of former unaccompanied refugee children 3–4 years after their return to Vietnam', *Child Abuse and Neglect*, 25: 249–263.

Lutkehaus, N.C. (1999) 'Missionary maternalism: gendered images of the Holy Spirit Sisters in colonial New Guinea', in Huber, M.T. and Lutkehaus, N.C. (eds) *Gendered missions: women and men in missionary discourse and practice*, Ann Arbor, MI: University of Michigan Press, pp. 207–235.

Lyby, E. (2001) 'Vocational training for refugees: a case study from Tanzania', in Crisp, J., Talbot, C. and Cipollone, D.B. (eds) *Learning for a future: refugee education in developing countries*, Geneva: UNHCR, pp. 217–259.

McGregor, J.A., Copestake, J.G. and Wood, G.D. (2000) 'The inter-generational bargain: an introduction', *Journal of International Development*, 12: 447–451.

McIntyre, P. (2002) *Putting children in the right: guidelines for journalists and media professionals*, Brussels: International Federation of Journalists.

McKechnie, J. and Hobbs, S. (2002) 'Work by the young: the economic activity of school-aged children', in Tienda, M. and Wilson, W.J. (eds) *Youth in cities: a cross-national perspective*, Cambridge: Cambridge University Press, pp. 217–245.

McLaren, P. and Pinkney-Pastrana, J. (2001) 'Cuba, Yanquizacion, and the cult of Elian Gonzalez: a view from the "enlightened" states', *Qualitative Studies in Education*, 14: 201–219.

McLennan, J.D. (2000) 'To boil or not: drinking water for children in a periurban barrio', *Social Science and Medicine*, 51: 1211–1220.

McRobbie, A. (1978) 'Working class girls and the culture of femininity', in Centre for Contemporary Cultural Studies (ed.) *Women take issue*, Boston, MA: Routledge and Kegan Paul, pp. 96–108.

Machel, G. (2001) *The impact of war on children*, London: C. Hurst and Co.

Majekodunmi, B. (1999) *Protection in practice: the protection of children's rights in situations of armed conflict: UNICEF experience in Burundi*, Florence: UNICEF Innocenti Research Centre.

Malone, K. (2001) 'Children, youth and sustainable cities', *Local Environment*, 6: 5–12.

Marcus, R. (2001) 'Big finance, small casualties', *Children and Development*, 1: 1–3.

Martin, C.J. (1982) 'Education and consumption in Maragoli (Kenya) households' educational strategies', *Comparative Education*, 18: 139–155.

Martin, J. and Tajgman, D. (2002) *Eliminating the worst forms of child labour: a practical guide to ILO Convention No. 182*, Geneva: International Labour Office and Inter-Parliamentary Union.

Marx, K. (1976/1867) *Capital: a critique of political economy, Volume 1*, London: Penguin.

Marx, K. and Engels, F. (1998/1848) *The communist manifesto*, London: Verso.

Massey, D. (1998) 'The spatial construction of youth cultures', in Skelton, T. and Valentine, G. (eds) *Cool places: geographies of youth cultures*, London: Routledge.

Matthews, H. and Limb, M. (1999) 'Defining an agenda for the geography of children: review and prospect', *Progress in Human Geography*, 23: 61–90.

Matthews, H., Limb, M. and Taylor, M. (1999) 'Young people's participation and representation in society', *Geoforum*, 30: 135–144.

Maundeni, T. (2002) 'Seen but not heard? Focusing on the needs of children of divorced parents in Gaberone and surrounding areas, Botswana', *Childhood*, 9: 277–302.

Mawson, A. (2000) 'Children, impunity and justice: some dilemmas from northern Uganda', in Carlson, L., Mackeson-Sandbach, M. and Allen, T. (eds) *Children in extreme situations: proceedings from the 1998 Alistair Berkley Memorial Lecture*, London: LSE Development Studies Institute, pp. 86–98.

Mayall, B. (1994) 'Children in action at home and school', in Mayall, B. (ed.) *Children's childhoods: observed and experienced*, London: Falmer, pp. 114–127.

Mayo, M. (2001) 'Children's and young people's participation in development in the South and urban regeneration in the North', *Progress in Development Studies*, 1: 279–293.

Mayo, P. (1995) 'Critical literacy and emancipatory politics: the work of Paulo Freire', *International Journal of Educational Development*, 15: 363–379.

Mead, M. (1928) *Coming of age in Samoa: a study of adolescence and sex in primitive societies*, London: Cape.

Mead, M. (1966) 'A cultural anthropologist's approach to maternal deprivation', *Deprivation of maternal care: a reassessment of its effects*, New York: Schocken, pp. 45–62.

Mehrotra, S., Vandemoortele, J. and Delamonica, E. (2000) *Basic services for all?*, Florence: UNICEF Innocenti Research Centre.

Messkoub, M. (1992) 'Deprivation and structural adjustment', in Wuyts, M., Mackintosh, M. and Hewitt, T. (eds) *Development policy and public action*, Oxford: Oxford University Press/Open University, pp. 175–198.

Milaat, W.A., Ghabrah, T.M., Al-Bar, H.M.S., Abalkhail, B.A. and Kordy, M.N. (2001) 'Population-based survey of childhood disability in Eastern Jeddah using the ten questions tool', *Disability and Rehabilitation*, 23: 199–203.

Milanovic, B. (1999) *True world income distribution, 1988 and 1993: first calculation based on household surveys alone*, Washington: World Bank.

Miles, S. (1996) 'Engaging with the disability rights movement: the experience of community-based rehabilitation in southern Africa', *Disability and Society*, 11: 501–517.

Miljeteig, P. (1999) 'Introduction: understanding child labour', *Childhood*, 6: 5–12.

Modell, J. (2000) 'How may children's development be seen historically?', *Childhood*, 7: 81–106.

Molyneux, M. (1985) 'Mobilization without emancipation? Women's interests, state and revolution in Nicaragua', *Feminist Studies*, 11.

Montgomery, H. (2000) 'Abandonment and child prostitution in a Thai slum community', in Panter-Brick, C. and Smith, M.T. (eds) *Abandoned children*, Cambridge: Cambridge University Press, pp. 182–198.

Montgomery, H. (2001) *Modern Babylon? Prostituting children in Thailand*, Oxford: Berghahn Books.

Moore, K. (2000) 'Supporting children in their working lives: obstacles and opportunities within the international policy environment', *Journal of International Development*, 12: 531–548.

Mungazi, D.A. (1988) 'Educational policy for Africans and church–state conflict during the Rhodesia Front government in Zimbabwe', *National Social Science Conference*, 3–5/11/88, Philadelphia, PA.

Muniz, J.G., Fabian, J.C. and Manriquez, J.C. (1984) 'The recent worldwide economic crisis and the welfare of children: the case of Cuba', *World Development*, 12: 247–260.

Murakami, H., Kobyashi, M., Zhu, X., Li, Y., Wakai, S., and Chiba, Y. (2003) 'Risk of transmission of hepatitis B virus through childhood immunization in northwestern China', *Social Science and Medicine*, 5: 1821–1832.

Myers, W.E. (1999) 'Considering child labour: changing terms, issues and actors at the international level', *Childhood*, 6: 13–26.

Myrstad, G. (1999) 'What can trade unions do to combat child labour?', *Childhood*, 6: 75–88.

Narman, A. (1995) 'Fighting fire with petrol: how to counter social ills in Africa with economic structural adjustment', in Simon, D., Van Spengen, W., Dixon, C. and Narman, A. (eds) *Structurally adjusted Africa: poverty, debt and basic needs*, London: Pluto, pp. 45–56.

Nelson, N. and Wright, S. (1995) 'Participation and power', in Nelson, N. and Wright, S. (eds) *Power and participatory development*, London: Intermediate Technology Publications, pp. 1–18.

Nieuwenhuys, O. (1994) *Children's lifeworlds: gender, welfare and labour in the developing world*, London: Routledge.

Nieuwenhuys, O. (1996) 'The paradox of child labor and anthropology', *Annual Review of Anthropology*, 25: 237–251.

Nordholm, L.A. and Lundgren-Lindquist, B. (1999) 'Community-based rehabilitation in Moshupa village, Botswana', *Disability and Rehabilitation*, 21: 515–521.

Nsamenang, A.B. (2002) 'Adolescence in sub-Saharan Africa: an image constructed from Africa's triple inheritance', in Brown, B.B., Larson, R.W. and Saraswathi, T.S. (eds) *The world's youth: adolescence in eight regions of the globe*, Cambridge: Cambridge University Press, pp. 61–104.

Nyambedha, E.O., Wandibba, S. and Aagaard-Hansen, J. (2003) 'Changing patterns of orphan care due to the HIV epidemic in western Kenya', *Social Science and Medicine*, 57: 301–311.

Obasi, E. (1997) 'Structural adjustment and gender access to education in Nigeria', *Gender and Education*, 9: 161–178.

Office of the High Commission for Human Rights (1993) 'Fact sheet no. 10 (rev. 1): The Rights of the Child', UNHCHR, accessed 07/05/02, http://www. unhchr.ch/html/menu6/2/fs10.htm.

O'Higgins, N. (2001) *Youth unemployment and employment policy*, Geneva: International Labour Office.

Oke, M., Khattar, A., Pant, P. and Saraswati, T.S. (1999) 'A profile of children's play in urban India', *Childhood*, 6: 207–219.

Otunnu, O. (2000) 'The Convention and children in situations of armed conflict', in Carlson, L., Mackeson-Sandbach, M. and Allen, T. (eds) *Children in extreme situations: proceedings from the 1998 Alistair Berkley Memorial Lecture*, London: LSE Development Studies Institute, pp. 48–57.

Pal, D.K., Chaudhury, G., Sengupta, S. and Das, T. (2002) 'Social integration of children with epilepsy in rural India', *Social Science and Medicine*, 54: 1867–1874.

Parker, F. (1960) *African development and education in Southern Rhodesia*, Columbus, OH: Ohio State University Press.

Parsons, T. (1951) *The social system*, London: Routledge and Kegan Paul.

Patrinos, H.A. and Psacharopoulos, G. (1995) 'Educational performance and child labor in Paraguay', *International Journal of Educational Development*, 15: 47–60.

Peil, M. (1995) 'Ghanaian education as seen from an Accra suburb', *International Journal of Educational Development*, 15: 289–305.

Penn, H. (2002) 'The World Bank's view of early childhood', *Childhood*, 9: 118–132.

Pfeiffer, J., Gloyd, S. and Li, L.R. (2001) 'Intrahousehold resource allocation and child growth in Mozambique', *Social Science and Medicine*, 53: 83–97.

Pflug, B. (2002) *An overview of child domestic workers in Asia*, ILO, Geneva, accessed 07/09/04, http://www.ilo.org/public/english/standards/ipec/publ/childdomestic/overview_child.pdf.

Piaget, J. (1972) *The principles of genetic epistemology*, London: Routledge and Kegan Paul.

Pillai, R.K., Williams, S.V., Glick, H.A., Polsky, D., Berlin, J.A. and Lowe, R.A. (2003) 'Factors affecting decisions to seek treatment for sick children in Kerala, India', *Social Science and Medicine*, 57: 783–790.

Pilotti, F.J. (1999) 'The historical development of the child welfare system in Latin America', *Childhood*, 6: 408–422.

Pineros, M., Rosselli, D. and Calderon, C. (1998) 'An epidemic of collective conversion and dissociation disorder in an indigenous group of Colombia: its relation to cultural change', *Social Science and Medicine*, 46: 1425–1428.

Pivnick, A. and Villegas, N. (2000) 'Resilience and risk: childhood and uncertainty in the AIDS epidemic', *Culture, Medicine and Psychiatry*, 24: 101–136.

Potter, R.B., Binns, T., Elliot, J.A. and Smith, D. (1999) *Geographies of development*, Harlow: Longman.

Pribilsky, J. (2001) '*Nervios* and "modern childhood": migration and shifting contexts of child life in the Ecuadorian Andes', *Childhood*, 8: 251–273.

Prochner, L. (2002) 'Preschool and playway in India', *Childhood*, 9: 435–453.

Prout, A. and James, A. (1990) 'A new paradigm for the sociology of childhood? Provenance, promise and problems', in James, A. and Prout, A. (eds) *Constructing and reconstructing childhood*, London: Falmer, pp. 7–34.

Psacharopoulos, G. (1981) 'Returns to education: an updated international comparison', *Comparative Education*, 17: 321–341.

Psacharopoulos, G. and Tzannatos, Z. (1991) 'Female labour force participation', in Psacharopoulos, G. (ed.) *Essays on poverty, equity and growth*, Oxford: Pergamon Press.

Punch, S. (2000) 'Children's strategies for creating playspaces: negotiating independence in rural Bolivia', in Holloway, S.L. and Valentine, G. (eds) *Children's geographies: playing, living, learning*, London: Routledge, pp. 48–62.

Punch, S. (2002a) 'Research with children: the same or different from research with adults?', *Childhood*, 9: 321–341.

Punch, S. (2002b) 'Youth transitions and interdependent adult–child relations in rural Bolivia', *Journal of Rural Studies*, 18: 123–133.

Punch, S. (2003) '"You'd better talk to my son because he knows more than me": primary education and youth transitions in rural Bolivia', *Uncertain transitions: youth in a comparative perspective*, presented at a conference at University of Edinburgh, 27/06/03.

Rabwoni, O. (2002) 'Reflections on youth and militarism in contemporary Africa', in de Waal, A. and Argenti, N. (eds) *Young Africa: realising the rights of children and youth*, Trenton NJ: Africa World Press, pp. 155–169.

Raffaelli, M., Koller, S.H., Reppold, C.T., Kuschick, M.B., Krum, F.M.B. and Bandeira, D.R. (2001) 'How do Brazilian street youth experience "the street"? Analysis of a sentence completions task', *Childhood*, 8: 396–415.

Rajbhandary, J., Hart, R. and Khatiwada, C. (2001) 'Extracts from *The children's clubs of Nepal: a democratic experiment*', *PLA Notes*, 42: 23–28.

Reynolds, P. (1995) 'Youth and the politics of culture in South Africa', in Stephens, S. (ed.) *Children and the politics of culture*, Princeton, NJ: Princeton University Press.

Rigi, J. (2003) 'The conditions of post-Soviet dispossessed youth and work in Almaty, Kazakhstan', *Critique of Anthropology*, 23: 35–49.

Rizzini, I. and Dawes, A. (2001) 'Editorial: on cultural diversity and childhood adversity', *Childhood*, 8: 315–321.

Rizzini, I. and Lusk, M.W. (1995) 'Children in the streets: Latin America's lost generation', *Children and Youth Services Review*, 17: 391–400.

Robson, E. (1996) 'Working girls and boys: children's contributions to household survival in West Africa', *Geography*, 81: 403–407.

Robson, E. (2000) 'Invisible carers: young people in Zimbabwe's home-based healthcare', *Area*, 32: 59–70.

Robson, E. (2004) 'Children at work in rural northern Nigeria: patterns of age, space and gender', *Journal of Rural Studies*, 20: 193–210.

Robson, E. and Ansell, N. (2000) 'Young carers in southern Africa: exploring stories from Zimbabwean secondary school students', in Holloway, S.L. and Valentine, G. (eds) *Children's geographies: playing, living, learning*, London: Routledge, pp. 174–193.

Roche, J. (1999) 'Children: rights, participation and citizenship', *Childhood*, 6: 475–493.

Rosemberg, F. and Andrade, L.F. (1999) 'Ruthless rhetoric: child and youth prostitution in Brazil', *Childhood*, 6: 113–131.

Rostow, W.W. (1960) *The stages of economic growth: a non-communist manifesto*, London: Cambridge University Press.

Rwezaura, B. (1998) 'Competing "images" of childhood in the social and legal systems of contemporary sub-Saharan Africa', *International Journal of Law, Policy and the Family*, 12: 253–278.

Sacks, R.G. (1997) 'Commercial sex and the single girl: women's empowerment through economic development in Thailand', *Development in Practice*, 7: 424–427.

Saldanha, A. (2002) 'Music, space, identity: geographies of youth culture in Bangalore', *Cultural Studies*, 16: 337–350.

Santa Maria, M. (2002) 'Youth in Southeast Asia', in Brown, B.B., Larson, R.W. and Saraswathi, T.S. (eds) *The world's youth: adolescence in eight regions of the globe*, Cambridge: Cambridge University Press, pp. 171–206.

Sargent, C. and Harris, M. (1998) 'Bad boys and good girls: the implications of gender ideology for child health in Jamaica', in Scheper-Hughes, N. and Sargent, C. (eds) *Small wars: the cultural politics of childhood*, Berkeley, CA: University of California Press, pp. 202–228.

Save the Children (1995) *Focus on images*, London: Save the Children.

Save the Children (2001) *Joint Public Private Initiatives: meeting children's right to health?*, London: Save the Children UK.

Save the Children (2002a) *The New Partnership for Africa's Development: are African children the new capital for trade and development?*, Toronto: Save the Children (Canada).

Save the Children (2002b) *Shaping a country's future with children and young people: National Plans of Action for Children: involving children and young people in their development*, Toronto: Save the Children.

Schade-Poulsen, M. (1995) 'The power of love: rai music and youth in Algeria', in Amit-Talai, V. and Wulff, H. (eds) *Youth cultures: a cross-cultural perspective*, London: Routledge, pp. 81–113.

Scheper-Hughes, N. and Hoffman, D. (1998) 'Brazilian apartheid: street kids and the struggle for urban space', in Scheper-Hughes, N. and Sargent, C. (eds) *Small wars: the cultural politics of childhood*, Berkeley, CA: University of California Press, pp. 352–388.

Scheper-Hughes, N. and Sargent, C. (1998) 'Introduction: the cultural politics of childhood', in Scheper-Hughes, N. and Sargent, C. (eds) *Small wars: the cultural politics of childhood*, Berkeley, CA: University of California Press, pp. 1–33.

Schildkrout, E. (2002/1978) 'Age and gender in Hausa society: socio-economic roles of children in urban Kano', *Childhood*, 9: 344–368.

Schlegel, A. and Barry, H. (1991) *Adolescence: an anthropological inquiry*, New York: Free Press.

Schoepf, B.G. (2001) 'International AIDS research in anthropology: taking a critical perspective on the crisis', *Annual Review of Anthropology*, 30: 335–361.

Schultz, T.P. (1993a) 'Investments in the schooling and health of women and men: quantities and returns', *Journal of Human Resources*, 28: 694–733.

Schultz, T.P. (1993b) 'Returns to women's education', in King, E.M. and Hill, M.A. (eds) *Women's education in developing countries*, Baltimore, MD: Johns Hopkins University Press for the World Bank

Seabrook, J. (2001) *Children of other worlds: exploitation in the global market*, London: Pluto.

Selolwane, O.D. (2003) 'The future of democracy in Botswana: the role of youth', *Youth and the politics of generational conflict in Africa*, 24–25 April, Leiden: The Netherlands.

Shotton, J. (2002) 'How pedagogical changes can contribute to the quality of education in low-income countries', in Desai, V. and Potter, R.B. (eds) *The companion to development studies*, London: Arnold, pp. 414–418.

Silberschmidt, M. and Rasch, V. (2001) 'Adolescent girls, illegal abortions and "sugar-daddies" in Dare es Salaam: vulnerable victims and active social agents', *Social Science and Medicine*, 52: 1815–1826.

Simms, C., Rowson, M. and Peatty, S. (2001) *The bitterest pill of all: the collapse of Africa's health systems*, London: Medact/Save the Children UK.

Singhal, A. and Howard, W.S. (2003) *The children of Africa confront AIDS*, Athens, OH: Ohio University Press.

Smith, L.C. and Haddad, L. (2000) *Explaining child malnutrition in developing countries*, Washington, DC: International Food Policy Research Institute.

Smith, P.K. and Cowie, H. (1988) *Understanding children's development*, Oxford: Blackwell.

Soares, K.V.S., Blue, I., Cano, E. and Mari, J.D.J. (1998) 'Violent death in young people in the city of Sao Paulo, 1991–1993', *Health and Place*, 4: 195–198.

Spinks, C. (2002) 'Pentecostal Christianity and young Africans', in de Waal, A. and Argenti, N. (eds) *Young Africa: realising the rights of children and youth*, Trenton, NJ: Africa World Press, pp. 193–206.

Stephens, S. (1994) 'Children and environment: local worlds and global connections', *Childhood*, 2: 1–21.

Stephens, S. (ed.) (1995) *Children and the politics of culture*, Princeton, NJ: Princeton University Press.

Stevenson, H.W. and Zusho, A. (2002) 'Adolescence in China and Japan: adapting to a changing environment', in Brown, B.B., Larson, R.W. and Saraswathi, T.S. (eds) *The world's youth: adolescence in eight regions of the globe*, Cambridge: Cambridge University Press, pp. 141–170.

Swart-Kruger, J. and Chawla, L. (2002) '"We know something someone doesn't know": children speak out on local conditions in Johannesburg', *Environment and Urbanization*, 14: 85–96.

Swart-Kruger, J. and Richter, L.M. (1997) 'AIDS-related knowledge, attitudes and behaviour among South African street youth: reflections on power, sexuality and the autonomous self', *Social Science and Medicine*, 45: 957–966.

Temba, K. and de Waal, A. (2002) 'Implementing the Convention on the Rights of the Child in Africa', in de Waal, A. and Argenti, N. (eds) *Young Africa: realising the rights of children and youth*, Trenton, NJ: Africa World Press, pp. 207–232.

Terra de Souza, A.C., Peterson, K.E., Andrade, F.M.O., Gardner, J. and Ascerio, A. (2000) 'Circumstances of post-neonatal deaths in Ceara, Northeast Brazil: mothers' healthcare-seeking behaviours during their infants' fatal illness', *Social Science and Medicine*, 51: 1675–1693.

Thind, A. and Andersen, R. (2003) 'Respiratory illness in the Dominican Republic: what are the predictors for health services utilization of young children?', *Social Science and Medicine*, 56: 1173–1182.

Thomas, D., Lavy, V. and Strauss, J. (1996) 'Public policy and anthropometric outcomes in the Cote d'Ivoire', *Journal of Public Economics*, 61: 155–192.

Thorne, S. (1999) 'Missionary–imperial feminism', in Huber, M.T. and Lutkehaus, N.C. (eds) *Gendered missions: women and men in missionary discourse and practice*, Ann Arbor, MI: University of Michigan Press, pp. 39–66.

Tolfree, D. (1995) *Roofs and roots: the care of separated children in the developing world*, Aldershot, UK: Arena (for Save the Children).

Ukwuani, F.A. and Suchindran, C.M. (2003) 'Implications of women's work for child nutritional status in sub-Saharan Africa: a case study of Nigeria', *Social Science and Medicine*, 56: 2109–2121.

UNAIDS (1999) *A review of household and community responses to the HIV/AIDS epidemic in the rural areas of southern Africa*, Geneva: UNAIDS.

UNAIDS (2003) *AIDS epidemic update: December 2003*, Geneva: UNAIDS/WHO.

UNDP (2003) 'Human development indicators, 2003', accessed 20/05/04, http://hdr.undp.org/reports/global/2003/indicator/index.html.

UNESCO (2000a) *Education for all 2000 assessment: global synthesis*, Paris: International Consultative Forum on Education for All.

UNESCO (2000b) *Education for all 2000 assessment: statistical document*, Paris: International Consultative Forum on Education for All.

UNICEF (1964) *Children of the developing countries*, Cleveland, OH: Nelson.

UNICEF (1996) 'Fifty years for children', accessed 16/05/02, http://www.unicef.org/sowc96/50years.htm.

UNICEF (1999) *Children in jeopardy: the challenge of freeing poor nations from the shackles of debt*, New York: UNICEF.

UNICEF (2000) *Poverty reduction begins with children*, New York: UNICEF.

UNICEF (2001a) *Poverty and children: lessons of the 90s for Least Developed Countries*, New York: UNICEF.

UNICEF (2001b) *Progress since the World Summit for Children: a statistical review*, New York: UNICEF.

UNICEF (2002a) *Adolescence: a time that matters*, New York: UNICEF.

UNICEF (2002b) *Adult wars, child soldiers: voices of children involved in armed conflict in the East Asia and Pacific region*, Bangkok: UNICEF East Asia and Pacific Regional Office.

UNICEF (2002c) *The state of the World's children*, New York: UNICEF.

UNICEF (2003a) 'Child protection', accessed 21/07/03, http://www.unicef.org/media/media_9482.html.

UNICEF (2003b) *Children in institutions: the beginning of the end?*, Florence: Innocenti Research Centre.

UNICEF (2003c) 'Facts on children', accessed 21/07/03, http://www.unicef.org/media/media_fastfacts.html.

UNICEF (2003d) *The state of the World's children*, New York: UNICEF.

UNICEF (n.d.) 'Goals set by the World Summit on Children, 1990', accessed 07/09/04, http://www.unicef.org/wsc/goals.htm.

United Nations (1989) 'Convention on the Rights of the Child', accessed 15/06/03, http://www.unicef.org/crc/crc.htm.

United Nations (1996) *Resolution adopted by the General Assembly: World Programme of Action for Youth to the Year 2000 and Beyond*, New York: United Nations General Assembly.

United Nations (1998) *Letter dated 11 September 1998 from the Permanent Representative of Portugal to the United Nations addressed to the Secretary-General*, A/53/378, New York: United Nations General Assembly.

United Nations (2001a) *Implementation of the World Programme of Action for Youth to the Year 2000 and Beyond: Report of the Secretary-General*, A/56/180, New York: United Nations General Assembly.

United Nations (2001b) 'World Population Prospects population database', United Nations Population Division, accessed 09/05/02, www.un.org/esa/population/demobase.

United Nations (2002) *World Youth Report 2003*, E/CN.5/2003/4, New York: United Nations Economic and Social Council.

Urassa, M., Boerma, J.T., Ng'weshemi, J., Isingo, R., Schapink, D. and Kumogola, Y. (1997) 'Orphanhood, child fostering and the AIDS epidemic in rural Tanzania', *Health Transition Review*, 7: 141–153.

US Census Bureau (2004) 'IDB Summary Demographic Data for Somalia', accessed 25/5/04, http://www.census.gov/cgi-bin/ipc/idbsum?cty=SO.

USAID/UNICEF/UNAIDS (2002) *Children on the brink 2002: a joint report on orphan estimates and program strategies*, Washington: USAID/UNICEF/UNAIDS.

Valentine, G., Skelton, T. and Chambers, D. (1998) 'Cool places: an introduction to youth and youth cultures', in Skelton, T. and Valentine, G. (eds) *Cool places: geographies of youth cultures*, London: Routledge, pp. 1–32.

Van Bueren, G. (1996) 'The quiet revolution: children's rights in international law', in John, M. (ed.) *Children in charge, volume 1: the child's right to a fair hearing*, London: Jessica Kingsley Publishers.

Verma, S. and Saraswathi, T.S. (2002) 'Adolescence in India: street urchins or Silicon Valley millionaires?', in Brown, B.B., Larson, R.W. and Saraswathi, T.S. (eds) *The world's youth: adolescence in eight regions of the globe*, Cambridge: Cambridge University Press, pp. 105–140.

Wagstaff, A. (2003) 'Child health on a dollar a day: some tentative cross-country comparisons', *Social Science and Medicine*, 57: 1529–1538.

Wagstaff, A. and Watanabe, N. (2000) *Socioeconomic inequalities in child malnutrition in the developing world*, Washington: World Bank.

Webb, D. (1997) *HIV and AIDS in Africa*, London: Pluto.

Welti, C. (2002) 'Adolescents in Latin America: facing the future with skepticism', in Brown, B.B., Larson, R.W. and Saraswathi, T.S. (eds) *The world's youth: adolescence in eight regions of the globe*, Cambridge: Cambridge University Press, pp. 276–307.

West, A. (1999) 'Children's own research: street children and care in Britain and Bangladesh', *Childhood*, 6: 145–155.

White, B. (1999) 'Defining the intolerable: child work, global standards and cultural relativism', *Childhood*, 6: 133–144.

White, H., Leavy, J. and Masters, A. (2002) *Comparative perspectives on child poverty: a review of poverty measures*, London: Young Lives, Save the Children UK.

Whiteford, L.M. (1998) 'Children's health as accumulated capital: structural adjustment in the Dominican Republic and Cuba', in Scheper-Hughes, N. and Sargent, C. (eds) *Small wars: the cultural politics of childhood*, Berkeley, CA: University of California Press, pp. 186–201.

Whitworth, A. and Stephenson, R. (2002) 'Birth spacing, sibling rivalry and child mortality in India', *Social Science and Medicine*, 55: 2107–2119.

WHO (1947) *World Health Organisation Constitution*, Geneva: World Health Organisation.

WHO (1998a) *Reducing mortality from major killers of children*, accessed 15/06/03, http://www.who.int/inf-fs/en/fact178.html.

WHO (1998b) *The second decade: improving adolescent health and development*, WHO/FRH/ADH/98.18, Geneva: Adolescent Health and Development Programme, Family and Reproductive Health, World Health Organization.

WHO (1999) *The evolution of diarrhoeal and acute respiratory disease control at WHO*, Geneva: Department of Child and Adolescent Health and Development.

WHO (2000) *What about boys? A literature review on the health and development of adolescent boys*, WHO/FCH/CAH/00.7, Geneva: Department of Child and Adolescent Health and Development, World Health Organization.

WHO (2001a) *Global Alliance for Vaccines and Immunization (GAVI)*, accessed 07/09/04, http://www.who.int/mediacentre/factsheets/fs169/en/.

WHO (2001b) 'Major causes of death among children under five, worldwide, 2000', Child and Adolescent Health and Development, accessed 14/06/03, http://www.who.int/child-adolescent-health/OVERVIEW/CHILD_HEALTH/piechart1.jpg.

WHO (2002a) *Adolescent friendly health services*, WHO/FCH/CAH/02.14, Geneva: Department of Child and Adolescent Health and Development, World Health Organization.

WHO (2002b) *Growing in confidence: programming for adolescent health and development*, WHO/FCH/CAH/02.13, Geneva: Department of Child and Adolescent Health and Development, World Health Organization.

WHO (2002c) *Strategic directions for improving the health and development of children and adolescents*, WHO/FCH/CAH/02.21, Geneva: Department of Child and Adolescent Health and Development, World Health Organization.

WHO (2003) 'Protein-energy malnutrition', accessed 07/06/03, http://www.who.int/nut/pem.htm.

WHO/UNFPA/UNICEF (1997) *Action for adolescent health: towards a common agenda*, WHO/FRH/ADH/97.9, Geneva: Adolescent Health and Development Programme, Family and Reproductive Health, World Health Organization.

WHO/World Bank (2001) *Better health for poor children: a special report*, Geneva: WHO/World Bank Working Group on Child Health and Poverty.

Wilkinson, J. (2000a) *Children and participation: research, monitoring and evaluation with children and young people*, London: Save the Children.

Wilkinson, J. (2000b) *What is child poverty? Facts, measurements and conclusions*, London: Save the Children.

Willis, B.M. and Levy, B.S. (2002) 'Child prostitution: global health burden, research needs, and interventions', *The Lancet*, 359: 1417–1422.

Woodhead, M. (1999) 'Combatting child labour: listen to what the children say', *Childhood*, 6: 27–49.

World Bank (1999) *World Development Report 1998/99: knowledge for development*, Oxford: Oxford University Press.

World Bank (2002) *Global economic prospects and the developing countries, 2002*, Washington, DC: World Bank.

World Bank (2003) 'Multi-country reports by HNP indicators on socio-economic inequalities', accessed 18/09/03, http://www.worldbank.org/poverty/health/data/statusind.htm.

Wulff, H. (1995) 'Introducing youth culture in its own right: the state of the art and new possibilities', in Amit-Talai, V. and Wulff, H. (eds) *Youth cultures: a cross-cultural perspective*, London: Routledge, pp. 1–18.

Young, L. (2004a) 'The "place" of street children in Kampala, Uganda: marginalisation, resistance and acceptance in the urban environment', *Environment and Planning D: Society and Space*, 21: 607–627.

Young, L. (2004b) 'Journeys to the street: the complex migration geographies of Ugandan street children', *Geoforum*, 35: 471–488.

Young, L. and Ansell, N. (2003a) 'Fluid households, complex families: the impacts of children's migration as a response to HIV/AIDS in southern Africa', *Professional Geographer*, 55: 464–479.

Young, L. and Ansell, N. (2003b) 'Young AIDS migrants in southern Africa: policy implications for empowering children', *AIDS Care*, 15: 337–345.

Yun, H. A. (2000) 'Trilogy of categorical unfitness: women, children and the new man against the state in Singapore', *Childhood*, 7: 61–79.

Zimbabwe (1982) 'Zimbabwe: review of education 1959–1979: current problems and prospects for the next 20 years', *Minedaf V: Conference of Ministers of Education and those responsible for economic planning of African member states on education policy and cooperation*, 28/6–3/7/82, Harare: Ministry of Education.

Zvobgo, R. (1986) *Transforming Education: the Zimbabwean experience*, Harare: College Press.

Index